I0120901

The Geography of the Everyday

The Geography of the Everyday Toward an Understanding of the Given Rob Sullivan

The University of Georgia Press Athens

Paperback edition, 2020
© 2017 by the University of Georgia Press
Athens, Georgia 30602
www.ugapress.org
All rights reserved
Set in 10.25/13.5 Minion Pro by Kaelin Chappell Broaddus

Printed digitally

Library of Congress Cataloging-in-Publication Data

Names: Sullivan, Robert E., 1952– author.
Title: The geography of the everyday : toward an understanding of the given /
    by Rob Sullivan.
Description: Athens : University of Georgia Press, [2017] |
    Includes bibliographical references.
Identifiers: LCCN 2017019751| ISBN 9780820351681 (hardback : alk. paper) |
    ISBN 9780820351674 (pbk. : alk. paper) | ISBN 9780820351667 (ebook)
Subjects: LCSH: Geography—Philosophy. | Human geography—Philosophy.
Classification: LCC G70 .S95 2017 | DDC 910/.01—dc23
    LC record available at https://lccn.loc.gov/2017019751

ISBN 9780820351674 (paperback : alk. paper)

CONTENTS

ACKNOWLEDGMENTS

I want to thank John Agnew, Michael Curry, Lisa Kim Davis, and John Mc-Cumber for their guidance while I was a graduate student at UCLA, working on "The Geography of the Everyday" as my dissertation. I would like to thank Bethany Snead, Jane Curran, Thomas Roche, and especially Mick Gusinde-Duffy of the University of Georgia Press for supporting this project. Finally, I want to thank the anonymous reviewers, whose criticism and counsel contributed much to this book.

INTRODUCTION

Some things are taken for granted.

In much of Asia, people use chopsticks.

In many regions of Africa, people use their fingers, eating out of a common bowl.

In what is called the West, people use a knife, a fork, and a spoon.

These habitual ways of eating are everyday practices. They are given, assumed, unquestioned. In fact, they are so given that they disappear into the background. And yet what is background for one person may be completely alien to another. Many from Europe and the United States would find eating with their fingers from a common bowl to be bizarre and unappealing, and many an African would find eating without the use of fingers to be just as bizarre and unappealing, while many a person from Asia would think the combination of a fork, a knife, and a spoon to be unwieldy when a pair of chopsticks would do the job just fine.

Another example. In the United Kingdom, vehicles travel on the left side of the road. In most of the rest of the world, they travel on the right side. When a driver from out of country arrives in the UK and attempts to drive on the "wrong" side of the road, panic and disorientation can ensue.

The everyday's givenness, along with its very capacity to fade into the background, make the everyday one of the trickiest elements of reality to analyze. How does the everyday construct itself? Can the everyday undo itself? In other words, can it become un-everyday? Or can something that is atypical *become* everyday? The everyday does change, sometimes slowly, sometimes even glacially, but change it does. For instance, same-sex marriage, while still perhaps not ordinary, is no longer so extraordinary that it is totally outside of the norm. In fact, as defined by law in many jurisdictions, it is well within the norms. Or a woman wearing pants is very much part of the given in this, the first quarter of the twenty-first century. But if we suddenly landed in the first quarter of the twentieth century, it would be very weird. So everyday mores, customs, and habits do change, though sometimes at a snail's pace, and as they change the fabric of what is given changes with them.

What if you went to a nightclub and someone had homework and books spread out on the dance floor and was trying to study? Or, conversely, what if you went to a library and some of the patrons were dancing as if they were in a nightclub? These are the structural elements that the given and the everyday introduce into the world. Places are defined functionally, and the given follows after. This is perhaps most clearly visible in the structure of a house. Most of us do not cook dinner in our bedrooms, entertain guests in our bathroom, or sleep in our kitchens. In fact, if we engaged in such behavior, people would consider us disoriented: literally, out of orientation or out of place.

The given, then, helps us gear into the world in prearranged ways, allowing places to be constructed functionally so that certain actions can easily take place within them. We can challenge those prearrangements by engaging in behavior that is not everyday for the place in which we find ourselves. Practicing trombone at a lecture or screaming in a public bathroom are good examples. Yet the consequences of engaging in such behavior can be severe and even brutal. And, typically, we never think of such options, let alone engage in them, because we are set so solidly into the fabric of the everyday that anything outside of its bounds is not even considered.

You could say that the everyday is life or reality or that which is, but that might be casting too wide a net over the everyday. Accidents and insanity and crime are all part of life and reality, but they are not part of the everyday, that is, unless one is an emergency room nurse, a mental health care professional, a criminal, or a police officer. Yet the everyday does cover a lot of ground, if one thinks of it as normal behavior or as mundane reality or as places fulfilling their functions.

So why a book about the everyday?

And why one about the *geography* of the everyday?

It is my claim that geography is precisely what is needed to *locate* the everyday. Without a location, the everyday is always floating, never nailed down. Without geography and its primary component, place, nothing can be *in place*, and therefore everything by necessity escapes. And though there has been much written about the everyday from sociologists (for instance, Erving Goffman's *The Presentation of Self in Everyday Life*), anthropologists (such as *An Anthropology of Everyday Life* by Edward T. Hall), feminists (the Everyday Feminism blog), psychologists (such as *The Psychology of Everyday Things* by Donald A. Norman), and others, no one has yet nailed down the concept and given it a foundation. It is my hope that this book will at least go some way toward that goal.

How do I expect to achieve such a goal?

In the process of examining the everyday through a geographical perspective, I assemble the essential elements by which to conceive of the everyday. Time, space, history, geography, reproduction, the body, and the geographical mind are examined in separate chapters and brought into what I am calling the astructural structure by which I am investigating the everyday.

How to begin such a task?

I'll start testing my idea that geography can lead to a more comprehensive understanding of the everyday by looking at Ian Hacking's essay "Between Michel Foucault and Erving Goffman: Between Discourse in the Abstract and Face-to-Face Interaction." In this essay, Hacking takes Goffman's microsocial approach as a fillip to instigate an analytical apparatus combining Goffman with Foucault's macrosocial approach: "'Between Foucault and Goffman': that suggests a middle ground between the French philosopher and the American sociologist. That would in turn imply that the two stand in opposition. Not so: they are complementary. One needs to stand between the two men in order to take advantage of both. There is a clear sense in which Foucault's analysis was 'top-down,' directed at entire 'systems of thought' ... Goffman's research was 'bottom-up'—always concerned with individuals in specific locations entering into or declining social relations with other people" (Hacking 2004, 277–278). The idea of using Foucault's analysis for a "panopticon view" upon the everyday while simultaneously using Goffman's for a ground-level view may be promising for a comprehensive analysis of the everyday.

Hacking points out some interesting parallels between this pair of thinkers, as well as some ways in which their theories might be melded into a greater whole. First, he notes that Goffman's work focuses on "what Goffman called total institutions: prisons, mental hospitals, concentration camps, monasteries, boarding schools, naval vessels" (Hacking 2004, 287). Foucault similarly studies what he calls "complete and austere institutions," but Goffman never undertook an analysis of the social or political assumptions underlying such institutions, whereas this is precisely the locus of Foucault's concern, that is, "the preconditions for and mutations between successive institutional forms" (Hacking 2004, 288). However, according to Hacking, Foucault pays little attention to the routines and practices by which such institutional forms are embedded in and then manifested within society. But this is precisely Goffman's terrain: he investigates the everyday "in rich detail," but, as opposed to Foucault, Goffman "gives no hint of how the surrounding structures themselves were constructed" and writes "nothing about the history of the social practices he described or the history of the total institution" (Hacking 2004, 288, 294). Thus, their complementary fit: Foucault probes the superstructural matrix of total institutions and their derivatives while Goffman prowls the halls of the same, jotting down notes about face-to-face interactions within such settings. Put them together, and one has circumscribed the institutional framework of society, from tip-top to bottom rock.

Again, a note of caution is warranted. If it seems too facile, it probably is. Besides, there are those who position Goffman and Foucault not as complementary but as congruent. In *Return of the Actor*, for instance, Alain Touraine slots Goffman and Foucault together (along with Marcuse, Althusser, and Bourdieu), for "beyond their differences," which are considerable, they share "a vision of the [social] system as order, and a conception of the actor as calculator and as player" (1988, 6). In his study of Goffman, Tom Burns does the same: "What Foucault and Goffman . . . have attempted is to construct a framework within which it might be possible to identify, observe and record the exercise of power in terms of . . . hegemony designed for 'normalisation'" (1992, 164). And Anthony Giddens contrasts and compares Goffman and Foucault vis-à-vis "their common concern with carceral organizations" and "their work" regarding "the positioning and disciplining of the body" (1984, 158, 157). So, contrary to Hacking's appraisal, some have noted the supplementary rather than the complementary fit of Goffman and Foucault.

But if we attempt a deeper analysis than Hacking provides, and especially if the currently unoccupied middle ground between Foucault and Goffman is properly filled in, Hacking's notion may lead to a more comprehensive understanding of the everyday, and perhaps to a systematic way to measure the approximate extent of the everyday as well.

But why should anyone be interested in Goffman anymore? He seems to have faded from view and was always considered somewhat marginal by sociologists, perhaps because his writing was a bit too aligned with a popular style, or perhaps because his work was also quite popular in the commercial sense, at least during the 1960s and 1970s. He draws upon an incredibly wide assortment of sources, from etiquette books to unpublished dissertations to excerpts from novels to peer-reviewed articles. I still find him extremely useful regarding the everyday. His observations on the minutiae of the quotidian are relevant and illuminating. And his connection to Foucault also deserves attention.

So our question then becomes: if there is some "middle ground between the French philosopher [Foucault] and the American sociologist [Goffman]," upon which "one needs to stand between the two men in order to take advantage of both" (Hacking 2004, 277), what is that middle ground? What is its geography? What is its topography? Who or what is the mediator between the microsociology of Goffman and the macrosociology of Foucault?

Finding the middle ground between two such thinkers is no ordinary task. To reiterate, what I hope to achieve in such a synthesis is a greater understanding of the everyday. That which lies between these two must be brought on board to fill in the thick middle between Foucault and Goffman.

The first chapter of this book starts with an analysis of Goffman's conception of the situation, an idea that in Goffman's hands lacks any theoretical underpinning and assumes "our Anglo-American" society as its foundation. Though Goffman is adept at formulating analytical concepts to plumb the social intricacies of situations, he does not *situate* the situation. Thus, there is an amorphous quality to Goffman's situation. Despite his acute observational skills, there is a sense of the freely floating to his concept of the situation, as if it exists without consideration of time and place, history and geography.

I then examine Foucault and his conception of the milieu, an idea that centers on the balance between the limits of security and risk. And what we discover is that though Foucault does an excellent job with time and place, history and geography, he does not provide much in the way of the details of ordinary life. Finally, I compare Foucault and Goffman, working toward a conclusion regarding the possibility of combining these two theories into a whole.

The second chapter, "The SpaceTimePlace Thing," is an attempt to add a foundation to the situation and the milieu. Through separate analyses of time, space, and place, I set the situation and the milieu within physical constraints. I also suggest that space, time, and place are indissoluble elements, inextricably bound together, and should be treated as such. However, the main purpose of the chapter is to bail out situation and milieu from their deterritorialized, despatialized, and atemporal *situations* by giving them the aid of time, space, and place.

But even with these additions, the vertical element of time has not been properly dealt with, for time never exists in the moment but resounds with the depth of that which has led to *this* time. Through an exploration of the ruse as postulated by Michel de Certeau, I demonstrate in the third chapter, "Time Goes Vertical; Space Yields In," that history must be accounted for in any analysis of anything whatsoever, including, of course, the everyday. The vertical trajectory of history is then matched with the horizontal trajectory of Henri's Lefebvre's spatial secretion. I use Lefebvre's theorization of spatial secretion as my entry into his work. I do this via a survey of Nigeria and its oil network as presented by the geographer Michael Watts. I use one of de Certeau's tactics, the ruse, as a tool to understand the figure of Schweyck, a clown/trickster figure in Czech and German literature. In this way, I hope to bring history and geography into our discussion, with the ruse of de Certeau used to recruit history and time and the spatial secretion of Lefebvre used to recruit geography and space.

In any study of the everyday, the elemental requirements of everyday reproduction must be included, with eating, sleeping, the housing question, and the elimination of waste matter being necessary for everyday social reproduction. Chapter 4, "What Marx Brought in from the Cold: Reproduction," examines an idiosyncratic development in the reproduction of the reserve army of the working class, exploring circular migration in China and the role of urban villages in this migration. I also analyze the pressure being brought to bear on the middle class of the United States. U.S. middle-class status has destabilized and eroded over the past forty years while the middle classes of Brazil, China, and India have risen, especially during the last decade. Finally, I look at the extreme wealth of the upper class with a sketch of some recent developments in high-end housing in London and Cairo.

The body must be brought into this examination, as it is, of course, essential to any experience of the everyday. I draw on Maurice Merleau-Ponty's phenomenology for this. But the body must have its indexical coordinates, its here and now and back and front, as every body is indexically oriented to its own particular position in space. Finally, I need to develop a deeper understanding of the senses than that presuming there are five separate and distinct senses. Instead a multimodal and cross-modal sensory system is brought into play with "separate" senses infusing and informing one another.

With this reconceived notion of the body established, the final chapter, "Bringing in Geography," reveals the requirement of incorporating things into any formulation of the everyday. Without the vibrancy of stuff and the materiality of things, there can be no everyday. I end with a discussion of the mind, its essential unity with the world leading us to the idea of a neuroge-ography as a way (or perhaps *the* way) to understand the mind. *The Geogra-phy of the Everyday*, then, has branched out from Anglo-American society to the four corners of the globe and then, taking an interior turn, into the very structure of the mind. Hopefully, the book also demonstrates that plac-ing the everyday in a geographical context substantially reduces the elusive-ness of the everyday, as perhaps it is only elusive because it is seldom placed within the context of a place.

A remark or two is in order regarding my overall schema put forward in *The Geography of the Everyday*. Though it may seem quite structured, my goal is a kind of systemless system, one that is polymer, reversible as well as invertible, much as a Möbius strip.

This structure is intended to create a domain in which the imperatives of structuration do not rule out the transformations necessary for change. As William H. Sewell puts it in his *Logics of History*, "What is needed is a con-ceptual vocabulary that makes it possible to show how the ordinary opera-tions of structures can generate transformations" (2005, 140). While I agree with Sewell, this possibility of structural transformation can be made even more flexible by acknowledging that whatever laws govern transformation change, as do also the laws that govern structuration. "Not only do param-eters change in response to changes in systems of which they are a part, but the laws of transformation themselves change," state Richard Levins and Richard Lewontin in *The Dialectical Biologist* (1985, 277).

Marx's methodology claims much the same thing: the systematic laws of feudalism create the conditions for feudalism's transformation into capital-ism, while the laws of capitalism create the conditions for its destruction and the creation of communism. What this means is that possibilities for change can be built into systems, an architectonic principle that should be verifiable by the structure of the system itself. Can the structure mobilize the transcen-dence of its own structure, which may mean that it may nullify its own order and negate its own principles, possibly leading to its own destruction? That the structure of the ideas in this book adheres to such conditions is one of my intentions.

And now let us try to locate the everyday.

CHAPTER 1

Starting with Goffman and Ending with Foucault

In this chapter I examine Goffman's conception of the situation and Foucault's conception of the milieu as placeholders for everyday existence. Such an examination can illuminate Hacking's notion of the power and efficacy of a theoretical combination of a bottom-up Goffman and a top-down Foucault. It also allows us to begin to develop a geographical conception of the everyday, both in the way Goffman and Foucault fulfill the exigencies of geography and the way in which they leave them as mere contingencies.

Goffman's Place, Goffman's Time, Goffman's Problems

"The key to understanding his [Goffman's] ethnographies is to see them as ethnographies of concepts rather than of places. They are utopian ideal-types (such as the total institution) existing nowhere" (Manning 1992, 17). So this lack of a place ("existing nowhere") leaves a dangling question: where is the *where* in Goffman's work? Does place exist for him? Are we situated anywhere, somewhere, or nowhere? In short, where is Goffman's geography?

Typically, Goffman uses the term "situation" to describe the particular nexus under examination. But is the situation *situated*? In other words, does the situation exist in space, time, and place? Or is it purely abstract? And wouldn't an ethnography of a *concept* have to be purely abstract? But wait, doesn't an ethnography have to be concrete? Or at least be based on the concrete?

And what exactly is a situation anyway? In *Frame Analysis*, Goffman states, "I assume that definitions of a situation are built up in accordance with principles of organization—which govern events—at least social ones—and our subjective involvement in them: frame is the word I use to refer to such of these basic elements as I am able to identify" (1974, 10–11). Goffman usually frames place as a setting or a stage on which social beings or actors perform their roles, though he does come close to a nice definition of place while elaborating the qualities of what he terms sitedness: "Whenever an individual participates in an activity, he will be situated in regard to it, this entailing exposure over a given range to direct witness, and an opportunity, over much the same range, to acquire direct observations. These latter implications of 'sitedness,' in conjunction with his auditing capacities, generate a series of points beyond which he cannot obtain evidence as to what is going on. He will find barriers to his perception, a sort of *evidential boundary*" (1974, 215; italics, Goffman). Such a description, based on the limits of optical and auditory perceptual capacities and bounded by a series of points beyond which such evidence cannot be obtained, puts conditions of inclusion and exclusion on place. Place, then, is conditioned by what can and cannot be perceived, with a series of points demarcating the border separating these two domains. Except that place is never mentioned; instead, situation and sitedness are the preferred terms.

A somewhat similar definition is provided by Goffman in his aptly named essay "The Neglected Situation." Here, Goffman performs a lexical apologia as if from a sociological confessional: "Let us face what we have been offhand about: social situations" (Goffman 1972, 63). He then goes on to attempt to clear up the situation about situation, at least in its social manifestation: "I would define a social situation as an environment of mutual monitoring possibilities, anywhere within which an individual will find himself accessible to the naked senses of all others who are 'present,' and similarly find them accessible to him. According to this definition, a social situation arises whenever two or more individuals find themselves in one another's immediate presence, and it lasts until the next-to-last person leaves" (Goffman 1972, 63). Again, situation is tied together by the limits of accessibility to the naked senses, and so a sensorial border is created around its domain. Before moving on, the Foucauldian surveillance hue of that wonderfully dystopian phrase, "mutual monitoring possibilities," should also be noted, possibilities that must, according to Goffman, be characteristic of any social situation.

And so let me return to Goffman's assumptions regarding his "definition of a situation." This definition is "built up in accordance with principles of organization," those "which govern events—at least social ones—and our subjective involvement in them." Sitedness, I assume (as Goffman does not supply his own definition), equates with being situated in a situation. But, of course, the situation in which one is sited is abstract, because Goffman is concentrating on the structural components of the definition of the situation and not the definition of the situation itself! No wonder we sometimes feel as if we are navigating through nowhere when we read Goffman.

But if we examine Goffman's field notes, an argument can be made that the situations described in the notes are, indeed, situated, located, and placed, as it were. The field notes are taken from real incidents happening at actual times and in actual places and so do not proceed from a strictly conceptual apparatus. In other words, they are "grounded," in the sense that one knows that the places he references, such as a movie theater and or a cafeteria, are based on actual places and not simply categorical universals standing in for particulars.

Now, while this is the case for much of Goffman's work, frequently Goffman resorts to conceptual places that seem to hover above reality, in a kind of domain of hypothetical suspension. And even in *Asylums*, based as it is on a year's study in St. Elizabeth's, though the book "analyzes the experiences of inmates in a Washington mental institution," it "by extension" is applicable to "any institution in which the time and space of subordinates are carefully monitored and restricted," that is, any total institution (Manning 1992, 8). "The result is an ethnography that is less a study of a specific hospital and more an ethnography of the concept of the total institution itself" (Manning 1992, 9). Manning adds that "Goffman's major interest [in *Asylums*] was to understand the ramifications of an analytic framework; understanding inmate culture at St. Elizabeth's hospital was only a secondary consideration" (1992, 155). So, once again, the concept is primary, not the place.

A similar sense of practicing conceptual ethnographies bleeds into Goffman's hypothetical examples. For instance, while discussing the ramifications attendant to "the game of surveillance," Goffman notes, "After a block or two, and well before suspicion is likely to be aroused, one tail can turn off and radio a second to pick up the trail at the next intersection" (1969, 48). Where are these blocks? Where is this intersection? Being that they only exist for hypothetical purposes, that is, to illustrate a point, they exist nowhere, and indeed, to fulfill their purpose, they must not exist anywhere else but nowhere.

So what is the problem? Or, perhaps better, *where* is the problem? Maybe it is located (if I can be allowed to use that term in this context) in the realization that even Goffman's hypothetical places are more or less of "our own Anglo-American" variety, but they are seldom, if ever, acknowledged as such. So that even in the example cited above, we somehow know that this intersection and these blocks are taken from an imaginative repository specifically geared to America, or at the least to what used to be called the West.

But is there anything inherently wrong with that? Only that it presumes specificity without enunciating that presumption and then takes that specification as a universal. Let me cite a few examples from *Relations in Public* that may clarify this point.

> A boy taking his girl into an amusement park picture booth, alive to the fact that the booth is seen as a place where couples go to neck, elaborately goes to the change booth for change and holds the necessary quarter up by means of two fingers, so anyone present will see that his intended use of the booth is innocent. A man looking at a girly magazine in a store specializing in this commodity may be careful to leaf through the magazine at a rapid pace, giving the same amount of attention to each page, as if looking for a particular article or wanting to see what in general a magazine like this could be like. (Goffman 1971, 129–130)

While these examples are both hypothetical, or at least appear to be so, they are in fact tightly hinged to time and place, specifically the United States of a distinctly Anglo-American variety, circa 1955, as the references to girly magazines and amusement park picture booths seem to predate 1971, the publication date of *Relations in Public*. And if they seem dated relative to 1971, as I believe they do, then surely they are doubly or even trebly anachronistic now. But the main point is that Goffman uses these examples as exemplars, that is, as if they transcend both time and space, when they are distinctly situated temporally, spatially, and culturally as well, as who can doubt that these hypothetical people are of "our own Anglo-American variety"? So that, while making a pretense of being outside of time and space, of having universal status, Goffman's analysis is clearly embedded within both time and space and is, indeed, particular to a certain place at a certain time, the United States of America of the 1950s and 1960s.

But then why have I (as well as others) branded Goffman as both ahistorical and ageographical? If his terms of analysis are embedded in a certain time and place, how can they not be embedded in both history and geography? Perhaps the confusion arises from Goffman's assumption of a norm without an acknowledgment that he is assuming such a norm, so that "our own Anglo-American society" seems to serve as a stand-in for life itself, as lived anywhere and everywhere.

This is a bit unfair, as Goffman does admit in the preface to *Relations in Public* that the "*full* location of the practices considered in this volume" are "the English-speaking world, the Anglo-American community, West European nations, Protestant countries, Christian society, and the West," and he has confessed that "the reference unit, 'American society,' (which I use throughout [*Relations in Public*]) . . . is something of a conceptual scandal"

(1971, xv; italics, Goffman). In further delineating the scope of this "scandal," Goffman goes on to say: "I can with least lack of confidence make assertions about my 'own' cultural group, the one with which I have had the most first-hand experience, but I do not know what to call this grouping, what its full span or distribution is, how far back it goes in time, nor how these dimensions might have to be changed, according to the particular behavior under question" (1971, xv). What is this grouping? Who belongs to it? As a Jewish intellectual and a Canadian immigrant, is Goffman himself a member of this normative group? One is prone to assume (at least I am prone to assume) that this group and "our own Anglo-American society" are equivalent, yet Goffman specifically states that he does not know what to call this group. But then he uses precisely the phrase "our own Anglo-American society" in order to name the group under question. Goffman also confesses to his geographic and demographic ignorance about this group ("I do not know . . . what its full span or distribution is") as well as to his historical ignorance regarding the group ("I do not know . . . how far back it goes in time"). And he also confesses to a lack of knowledge regarding prognoses about the group's modes of behavior on both a geographical and a historical front: "I do not know . . . how these dimensions might have to be changed, according to the particular behaviors under question."

And so one is left with a distinctly sinking feeling in regard to both history and geography, or time and place, in terms of Goffman's work. On the one hand, things sometimes seem to be tied to particular epochs and specific locations; on the other hand, everything seems to be floating, eradicated conceptual ethnographies in a theoretical void. Assumptions that seem to necessarily depend on a reading implicating a standard brand, even stereotypical, white Protestant Americana culture circa 1955 are sealed with the imprimatur of the universal. "Given their social identities and setting, the participants will sense what sort of conduct *ought* to be maintained as the appropriate thing, however much they may despair of it actually occurring," Goffman states in "Embarrassment and Social Organization" (1956, 268; italics, Goffman). Yet this is exactly the problem, as such things as "social identities and settings" are not given but come with their burdens of history and of geography, and, furthermore, they are formed following the rules and codes of specific social orders. What I am suggesting is that some temporal and spatial discipline must be applied to Goffman's work to make it whole.

But there's another problem with Goffman's oeuvre as well. This has to do with the disconnection, or the lack of any attempt to even make a connection, between the tightly circumscribed and highly ritualized interaction order that Goffman so deftly delineates and the larger apparatus of the social world from which such elements derive. Or, rather, from which we assume they derive, this qualifier inserted because Goffman generally offers no rationale as to how such elements are created or whether their derivations can be traced to any sort of provenance, as they seem to appear fully formed from out of a sociological blue. As Tom Burns describes this problem, "There is hardly any discussion in his [Goffman's] writings of the way in which the traffic of social interaction, which is the stuff of social order, organises itself, or is organised, so as to constitute society, which we ordinarily conceive of as populated by organisations and social institutions, large and small . . . ranging from the U.S. Congress to the corner shop" (1992, 359).

This may be a bit overboard, as, of course, Goffman does place total institutions, such as mental hospitals, convents, orphanages, and the military, within his ambit, but it may not be too much of an exaggeration to claim that Goffman does not pay sufficient attention to how numerous iterations of particular types of social interactions mount up to "the stuff of social order." But even here, Goffman does at least present *some* evidence for the processes that constitute such structural formations. For instance, in *Asylums*, Goffman quotes Ivan Belknap's *Human Problems of a State Mental Hospital* to illustrate the pedagogical functions of punishment and humiliation: "This knowledge [of shock therapy among the patients] is based on the fact that some of the patients in Ward 30 have assisted the shock team in the administration of therapy to patients, holding them down, and helping to strap them in bed, or watching them after they have quieted" (Goffman 1961, 33).[1] Here at least the outlines of the dissemination of discipline can be discerned, from the mental hospital's administrators and physicians to the patients who assist in the spectacle of shock therapy to the patients in general. For the patients this "lesson" of "what may be done to them" is autodidactically absorbed: having seen what happens to those who "misbehave," they alter their behavior, squeezing it between normative parameters. Such a lesson can lead to what Goffman calls "conversion," depicted as a "mode of adaptation to the setting of a total institution" in which "the inmate appears to take over the official or staff view of himself and tries to act out the role of the perfect inmate . . . presenting himself as someone whose institutional enthusiasm is always at the disposal of the staff" (1961, 63). Of course, they must have also reported on the treatments to the other patients, thereby distributing the lesson to a wider audience. There is a trace of a genealogy hinted at here, but

Goffman never thoroughly follows up on this, especially when he is describing normative behavior in Anglo-American society. It's as if such behavior has existed for all time, its genealogical tracings and archeological remains not even mentioned, let alone inspected.

I am deliberately employing Foucauldian terms here, as what I want to suggest is that the Foucauldian methodologies of archeology and genealogy are precisely what is needed to fill out Goffman. He needs history, he needs time, he needs a delineation of the means by which interaction rituals are connected to social orders and institutions. And he needs space and place as well. Goffman himself was not unaware of these shortcomings, as he admits that he never theorized the roles that time and place or history and geography should play in a sociological analysis. In his address to the American Sociological Association in 1982 (the year of his death), Goffman states, "Orderliness is predicated on a large base of shared cognitive presuppositions, if not normative ones, and self-sustained restraints. How a given set of such understandings comes into being historically, spreads and contracts in geographical distribution over time, and how at any one place and time particular individuals acquire these understandings are good questions, but not ones I can address" (1983, 5). I take this to mean that he not only cannot address these issues in this, the presidential address he is delivering, but also that he cannot address them at all, given his focus on microanalysis.

Goffman also needs a more deliberative midlevel analysis of how individual selves and social groups are interrelated, or of how the macro and the micro scales of social analysis can be linked at the meso level so the full range of the social order can be illuminated. But, once again, Goffman was aware of this shortcoming, as during the same address he states, "Insofar as agents of social organizations of any scale, from states to households, can be persuaded, cajoled, flattered, intimidated, or otherwise influenced by effects only achievable in face-to-face dealings, then here, too, the interaction order impinges on macroscopic entities" (1983, 8). Here, Goffman does not so much analyze this midpoint meeting place between the macro and the micro but simply thrusts his way into that realm, shoving his analysis into the breach. In other words, he never truly accomplishes such an analysis in his work but seems to be attempting to patch over this hole by bodily tossing his form of interaction analysis into a vacuum.

Prior to taking a brief respite from Goffman, let me once more iterate that he has set forth a standard that can still stand as the criteria par excellence by which to analyze social interactions through the most minutely performed observations of the slightest bits of behavior, what Giddens, speaking specifically of Goffman, calls "the normatively regulated control of what might seem . . . to be the tiniest, most insignificant details of bodily movement or expression" (1984, 78) that is performed on a "never-to-be-relaxed—monitoring of behavior and its contexts" (Giddens 1990, 37).

Goffman took as his field the "face-to-face domain," and he has left us with a rich legacy, a legacy that can be used as the first step on the first level of an overall analysis of the everyday (Goffman 1983, 2). The other thing that Goffman has bestowed upon us, albeit it in a negative mode or as a lesson-in-reverse, as it were, is that his narrow construal of "our own Anglo-American society" needs to be widened out so that it includes history and geography as well as other social and ethnic groups, tasks I progressively assume as I continue this investigation.

Foucault's Milieu

Let me now turn to Foucault's conception of the milieu. Foucault's milieu is the ground upon which he positions the relationship between security and risk. Each milieu possesses its own dynamic between these two poles, a dynamic balanced between the limits of security and risk, a sense of equipoise that is sometimes stable, sometimes teetering, and at other times falling into either the mayhem of risk or the ossification of security. "The space in which a series of uncertain elements unfold is, I think, roughly what one calls the milieu," states Foucault in *Security, Territory, Population*, the compilation of his lectures at the Collège de France in 1977–78 (2007b, 20). "What is the milieu? It is what is needed for action at a distance of one body on another. It is therefore the medium of an action and the element in which it circulates. It is therefore the problem of circulation and causality that is at stake in this notion of milieu" (Foucault 2007b, 20–21).

There is, then, a striking difference between Goffman's situation and Foucault's milieu. The situation à la Goffman is close at hand, intimate, personal, always occurring within the physical propinquity of bodies with one another and in one place, whether that place be an office, a party, an elevator, or a bus. Foucault's milieu is what animates as well as what controls the situation, but as it is animated and controlled on a distal basis. It is impersonal, distant, alien, yet present through the power of its reach, a reach instigated both by technology and discipline. The milieu is that ground on which a homeostatic tension exists between the abyss of risk and the cove of security.

Here we should also insert a brief exegesis of three other influential uses of the term "milieu" by French thinkers, one descending from the geographer Vidal de la Blanche, one from the critic and historian Hippolyte Taine, and the last from Gilles Deleuze and Félix Guattari. In de la Blanche, "society and milieu" serve as stand-ins for "sociality and territoriality" (Buttimer 1971, 3). These two elements are construed as the "two scaffolds of human experience" that "interact to produce the earth's variegated cultural landscapes: society a complex web of organizational arrangements, milieu a variegated mosaic of physically differentiated regions" (Buttimer 1971, 3). So this conception is essentially physical and rooted in the region, the primary object of research of "the 'first generation' of French human geographers in their regional monographs" (Buttimer 1971, 3n4). Taine's trio of race, milieu, and moment had an enormous impact on both the writing of fiction and literary criticism, as Zola and de Maupassant were deeply indebted to this conceptual schema as they researched various environments in which to set their works (historicism can also be traced back to Taine's construal of the triad of race, milieu, and moment). In their rather arcane metaphysical toxology, Deleuze and Guattari use "milieu" to designate "the material field in which strata and assemblages are formed," and so the milieu as conceptualized by Deleuze and Guattari seems to retain the physical element bestowed upon it by French geographers of the region (Bonta and Protevi 2004, 113). However, that would be to underestimate Deleuze and Guattari's uncanny ability to multiply simplicity into complexity: "A milieu is a 'soup' or the coded medium of particle-flows and the strata that gives birth to or at least supports a rhizomatic assemblage: a living being, a symbiont, or an ecosystem, for example" (Bonta and Protevi 2004, 113). This codicil appears to push Deleuze and Guattari's construal a step or two closer to Foucault's, though Foucault seems to want to keep the term more closely aligned with an event or a series of events within an assemblage; in other words, Foucault's milieu is not a medium, it is that which occurs within the medium, and a distinctly human, not a natural, medium: the word "ecosystem" would ring false if it were contained within a Foucauldian analysis of the milieu. In "Notes on the Translation and Acknowledgments" to Deleuze and Guattari's A Thousand Plateaus, Brian Massumi states, "In French, milieu means 'surroundings,' 'medium,' (as in chemistry) and 'middle'" (Massumi 1988, xvii; italics, Massumi). He then goes on to add, "In the philosophy of Deleuze and Guattari, 'milieu' should be read as a technical term combining all three meanings" (Massumi 1988, xvii).

Foucault's construal of the milieu, as I understand it, can be of multiple types: political conventions, academic conferences, film locations, sporting events, church services, assisted living facilities, homeless encampments, business meetings, and so on. Notice that these disparate types share the quality of what could be called, albeit somewhat clumsily, "eventful occurrences." Foucault says, "Security will try to plan a milieu in terms of events or series of events or possible elements, of series that will have to be regulated within a multivalent and transformable framework. The specific space of security refers then to a series of possible events; it refers to the temporal and the uncertain, which have to be inserted within a given space" (2007b, 20).

To illustrate this, one might imagine the planning of a meeting of the leaders of the G-20. Uncertain contingencies (i.e., risks) have to be imagined and then inserted as possible givens into the space of the meeting. It is only then that the responses of security forces to such risks can be developed and prefigured into the (possible) future of the milieu. Knowledge of the typology of the milieu as well as the specifications of a particular milieu, both on the part of those attempting to secure the milieu and on those willing to risk themselves in order to disrupt or even destroy the milieu, allows each side to perform at an optimum level. In other words, each specific milieu provides a more or less known platform from which to calibrate the possibilities of the limits of risk and security within that milieu. This is why there are "officials of the French electrical utility . . . [who] have at their disposal *catastrophe simulators*" to plan for possible nuclear accidents; such simulators are "comparable to the army's 'strategic calculators,' and akin to airplane and automobile simulators," in that they attempt to delimit the risk of that which Paul Virilio insists is inherent to the very processes/machines so delimited (Virilio 1993, 216).

These considerations are also of utmost importance during any civil disturbance or protest action. For instance, the civil rights demonstrators of the American South during the 1960s understood, to some degree, the parameters of risk and security of the milieu in which they were operating. In other words, they knew that beatings, jail, and possibly even death were the outer bounds to which risk could push security, and yet, despite the risks, they were willing to challenge those parameters. That is, their effort consisted of and constituted in and of itself a challenge to the very construction of the milieu in which they existed. On the other side, the police forces of the South were attempting to keep intact the preexisting milieu, which, of course, entailed *securing* it against the challenge of a reconfiguration of that very milieu by the protestors.

Foucault outlines the contours of the milieu as he delineates the problem of the circulation and control of population that arose in the latter half of the eighteenth century during the first wave of the Industrial Revolution. The increase in population meant that cities could no longer be regulated in a casual way as the walls of the cities became ineffectual against the tide of humanity passing into and out of the city. The city walls no longer could suppress the arrival of strangers, whether those strangers be merchants, mendicants, or murderers. This, then, constitutes the gist of the problem of population, as its circulation must somehow be controlled within the milieu of the modern city, a city now too expansive and too overflowing to curb the arrival and departure of an exploding population.

But circulation in relation to a burgeoning population also has a more physical dimension. The connotation here is to the circulation of air, water, and transportation systems. Circulation and centralization are tied together as the state attempts to extend its range of powers into the collective body. "Throughout the whole second half of the eighteenth century we see a huge effort being made to homogenize, normalize, classify, and centralize medical knowledge" (Foucault 2003, 181). The effort of minimizing the risk of contagion and disease while maximizing the centrality of information and the institutionalization and codification of all things medical also becomes the pivot point around which circulation, population, causality, and urban planning connect with the milieu and its double aspects of security and risk. "Security will rely on a number of material givens. It will, of course, work on site with the flows of water, islands, air, and so forth. . . . It is . . . a matter of maximizing the positive elements, for which one provides the best possible circulation, and minimizing what is risky and inconvenient, like theft and disease, while knowing that they will never be completely suppressed" (Foucault 2007b, 19). This passage includes an acknowledgment that security (and thus the milieu) has an outer boundary ("theft and disease . . . will never be completely suppressed") at which security and its techniques reach a limit. Beyond that limit, to which the algorithms of possibilities and probabilities cannot reach, lies the remainder, the residue of that which cannot be controlled, and which, then, is fertile ground for transformation and revolution, but also a miasma in which corruption of every kind and disease in every form can fester and breed.

The techniques of discipline frequently require "*enclosure*, the specification of a place heterogeneous to all others and closed in upon itself" (Foucault 1995, 141; italics, Foucault). Prisons, military barracks, monasteries, workshops and factories, boarding schools and colleges, hospitals and insane asylums, though organized for the control of certain segments of the population, work on those selected segments on an individual basis. "It does this first of all on the principle of elementary location or *partitioning*. Each individual has his own place; and each place its individual" (Foucault 1995, 143; italics, Foucault). Here what we have is "rule, the law, prohibition" marking "the limits between what is permitted and what is forbidden" (Foucault 2007a, 154). The techniques of the milieu, however, are not based on enclosure and the individual per se; instead, they are based on more or less open spaces and a population that comes and goes. Therefore, the milieu and its population, derived from a population that grew at such rapid rates during the 1700s and 1800s, open up an entirely new forum in which the forces of security and risk must learn to operate. Within this new forum, the techniques of the milieu must allow for the contingencies of the arrival of two unknowns, the stranger and the future. Since they are unknowns, and therefore essentially unknowable, ipso facto the techniques used to control them can only account for so much. Here again, they reach a limit, and beyond that a remainder exists in which security is no longer reliable and risk comes into play, a risk leveraged upon two unknowns, the future and strangers.

The techniques of the milieu must also operate at a distance. As its subjects are not confined and do not necessarily exist in the present (and so are not confined to either the here or to the here and now), these techniques must work across space and through time. In order for such a diffusive process to occur and at such a scope and scale, a great deal of centralization must occur. This necessitates the hub-and-network system congruent with the modern nation-state, with the centralizing power of the capital relaying signals of command and control to distant provinces and, in turn, receiving signals of compliance in the form of revenue and conscripts from the provinces. "Through a multitude of . . . mobile relays, relations are established between those who are spatially and temporally separated, and between events and decisions in spheres that none the less retain their formal autonomy" (Rose 1996, 43). For such actions at a distance to occur, transportation and communication networks have to be planned and constructed to extend out to the most remote areas of the state. And these must be extended while still maintaining the absolute centrality of the state's power; the more remote those areas, the more power the center must retain in order to keep what is at a distance in its grip: "dispersal occurs under conditions of concentration of control" (Sassen 2002, 15).

Let me reiterate that each milieu has a limit beyond which its security is at risk. This limit may be prefigured, at least to a certain extent, or it may alter as the event that it circumscribes develops. If the political protest is taken as a prototypical milieu, the workings of the milieu in all its forms, with its limits arranged around parameters defined by various understandings of security and risk, can perhaps be better understood. I proceed this way, despite what Foucault claims about the milieu, namely, that the regulation of the milieu entails "not so much establishing limits and frontiers, or fixing locations, as, above all and essentially, making possible, guaranteeing, and ensuring circulations: the circulation of people, merchandise, and air, etcetera" (2007b, 29). The reason I proceed with an emphasis on the limit is that even if one privileges circulation over limits, one still must master the limit of circulation in order to circulate whatever one intends to circulate, be that people, merchandise, or air. In other words, the guaranteeing of circulation, its assurance, relies, and relies necessarily, on the establishment of limits; ergo, we focus on limits, at least for now.

Of course, all of this does not mean that Foucault, when using the term, limits himself to his own construal of the milieu. For instance, in *The Birth of Biopolitics*, Foucault, discussing criminality, states, "penal action must act on the interplay of gains and losses or, in other words, on the environment; we must act on the market milieu in which the individual makes his supply of crime and encounters a positive or negative demand" (2004, 259). Similarly, Foucault refers to the "criminal milieu" and the "delinquent milieu" when discussing the genealogy of the prison system (Foucault 1980, 195–96). Here, milieu does not seem to be used to refer to either any form of circulation or a series of events, but rather to a limit circumscribing the limits of security and risk. In the same series of lectures, Foucault refers to "an economic and political choice formed and formulated . . . within the governmental milieu" (2004, 218). Here, the milieu seems even more static, as it is instantiated within a government and embedded within a state. Finally, in the same lecture (14 March 1979), Foucault, now with migration as his topic, mentions the "psychological cost for the individual establishing himself in a new milieu" (2004, 230). Here, "milieu" appears to be used as a synonym for home, niche, dwelling, or environment. There is no sense of a series or of circulation implied here, and although every home has its limit beyond which security is at risk, the use of the term in this specific case is so general as to be ideal, abstract, and formless. Granted, these examples are a far cry from the total sample of Foucault's use of the term; however, they do offer a limited amount of evidence that milieu was a wobbly term for Foucault, at least when used in his own discourse. But there is nothing odd about a thinker not being internally consistent with his concepts or not completely adhering to his own lexical decrees.

Foucault and Goffman:
Milieu and Situation,
Top-Down and Bottom-Up

It is rather difficult, if not impossible, to decipher Goffman's situation vis-à-vis Foucault's milieu through a strategy of contrasting and comparing them, as Foucault's notion of the milieu is overtheorized and underused (nearly to the point of being not used at all), while Goffman's notion of the situation is overused while being undertheorized (nearly to the point of not being theorized at all). So they are obverse mirrors of one another, both of them hypertrophic but not homologous.

One of Goffman's main preoccupations is the social situation, for he is continually dissecting its particulars. In the case of Foucault and the milieu, however, it is not so clear if the theorization of the milieu is of primary significance for him. It seems that it is not, as he does not—except for the few passages of *Security, Territory, Population* that I have cited—directly deal with the concept (at least from what I have been able to verify via my reading of his immense oeuvre). However, surely the milieu can be intercalated as a vital part of many of the subjects that Foucault investigates. One could fairly say that, for Foucault, there is the milieu of the prison, the milieu of the clinic, the milieu of the academy, and so on, and that with all of these Foucault is concerned with their equilibrium, their homeostasis, and their demarcations of security and risk, or what could be called the limits of their limits.

Goffman is also concerned with limits and homeostasis, though, of course, he doesn't use such terminology. He wants to demonstrate what normative behavior for Anglo-American society circa 1960 consists of, and so he must mark out its limits. However, where Foucault uses an abstract term, such as "the limit," to mark such a boundary, Goffman uses a concrete thing, such as a window, to mark such a boundary.

Perhaps if one thinks of Foucault's milieu as denoting types or universals and Goffman's situation as denoting tokens or particulars, one can hang Foucault's milieu and Goffman's situation together. A milieu, then, may be a prison or a clinic as a general category, whereas a situation takes places in *this* prison or *that* clinic as a specific place and during a particular period. Notice also that the milieu is static whereas the situation *occurs*. So maybe the milieu is a stand-in for the situation as a noun-like thing and the situation a stand-in for the milieu as a verb-like thing.

But then it isn't quite right to delimit Foucault in this manner, as Hacking does when he avers that Foucault "wrote of discourse in the abstract" as if that could stand as a categorical statement, with no possibility of garnering evidence to demonstrate that Foucault frequently trained his sights on the particular and the concrete (2004, 278). In fact, Foucault often cites particular prisons, clinics, and schools wherein specific events occurred. For instance, in *Discipline and Punish*, when Foucault wants to explicate the milieu of the public execution, he doesn't simply rattle on in generalities, dealing merely in abstract universals, but instead engages in concrete particularities: "To clarify the political problem posed by the intervention of the people in the spectacle of the executions, one need only cite two events. The first took place at Avignon at the end of the seventeenth century. It contained all the principle elements of the theatre of horror: the physical confrontation between the executioner and the condemned man, the reversal of the duel, the executioner pursued by the people, the condemned man saved by the ensuing riot and the violent inversion of the penal machinery" (1995, 63–64).

What Foucault is depicting in this passage is a breakdown of the milieu of the public execution, as the execution is inverted: the scene ends with the gallows smashed into pieces and hurled into the Rhone, the condemned man pardoned by an archbishop, and the executioner nearly torn to pieces by the mob. Foucault contrasts this breakdown of the homeostasis of the milieu of the execution with a "successful" execution that occurred in Paris in 1775: "Between the scaffold and the public, kept at a safe distance, two ranks of soldiers stood on guard, one facing the execution that was about to take place, the other facing the people in case of riot. Contact was broken: it was a public execution, but one in which the element of spectacle was neutralized, or rather reduced to abstract intimidation. Protected by force of arms, on an empty square, justice quietly did its work" (1995, 65).

The milieu of the public execution can only proceed if contact is broken between the scaffold and the people, that is, if the forces enforcing the proceedings tamp down the incipient violence of the crowd, reducing that sense of impending violence by "abstract intimidation," as Foucault puts it. Yet it will not be too long—a mere fourteen years—until the scene is not only inversed but reversed: the public will enforce the spectacle of the guillotine, and those hitherto in power will have their heads on the block. Of course this is simply a continuation of the milieu, merely reconfigured under a different cloak of authority, just as Stalin's rule is a continuation in reverse of the rule of the Tsars and the Soviet secret police, the NKVD, a continuation in reverse of the Tsar's secret police, the Okhranka. "Do you think it would be much better to have the prisoners operating the Panopticon apparatus and sitting in the central tower, instead of the guards?" (Foucault 1980, 164–65).

And it isn't quite right to delimit Goffman either, as he doesn't *always* confine his remarks about situations to those in situ. Goffman's situation isn't always a situation per se but can connote the situation in general as well. For instance, in *Behavior in Public Places*, Goffman states: "In many social situations, a particular main involvement will be seen as an intrinsic part of the social occasion in which the situation occurs, and will be defined as preferential if not obligatory. At a card party, for example, participants may be expected to focus their attention on cards, justifying their allocation of involvement by reference to the nature of the social occasion" (1963, 50). In this passage, we can notate quite different scales of generality: the all-inclusive and vaguely formulated social *situation*; the social *occasion*, nested within the social situation; and then, finally, the situation itself, nested within the social occasion, that is, the very situation we happen to be examining or observing "now." And Goffman parses the situation into types as well: "foreign students in a classroom," "persons not British at a cricket match," "when an individual in a vehicle sits or stands while awaiting his destination," "a woman not closely related to the deceased who appears at the funeral in a very modish, very complete, black ensemble," and so on (Goffman 1963, 50, 51).

Perhaps it is this very way in which Foucault generally confines himself to the abstract while Goffman generally confines himself to the concrete that leads Hacking to slot the former into the top-down and the latter into the bottom-up category. And, at first glance, especially in regard to slotting Foucault as one who confines himself to operating at the superstructural level, it is a somewhat understandable mistake. But a mistake it plainly is. The error becomes clear when one peruses Foucault's lectures and the various interviews conducted with him, which offer up numerous instances of Foucault slotting his own work into the bottom-up side of the ledger or his "continual referencing of *the local*" (Philo 2007, 358; italics, Philo).

For instance, in *Society Must Be Defended*, Foucault states that instead of deducing "whatever we like from the general phenomenon of the domination of the bourgeois class . . . we should be doing quite the opposite," or in other words: "We should be looking in historical terms, and from below, at how control mechanisms could come into play in terms of the exclusion of madness, or the repression and suppression of sexuality; at how these phenomena of repression or exclusion found their instruments and their logic, and met a certain number of needs at the actual level of the family and its immediate entourage, or in the cells or the lowest levels of society" (2003, 32). Foucault reiterates and expands upon his "bottom-up" approach in the same lecture (14 January 1976): "Power is exercised through networks, and individuals do not simply circulate in those networks; they are in a position to both submit to and exercise this power. They are never the inert or consenting targets of power; they are always its relays" (2003, 29).

Foucault also seems to believe that once rules are established, they are followed and followed in an absolute sense. But this is not necessarily so. Enormous gaps frequently exist between the de jure and the de facto state of affairs in terms of obedience to laws and codes, and rules and regulations. For instance, it is now illegal in the state of California to talk on a cell phone while driving. For a very brief period of time after the law was passed, it seemed to be in effect on a de jure basis: that is, the law was followed. After a few months, it was generally determined that the police were not about to enforce this regimen in anything approaching an absolute degree, and so drivers went back to driving while talking on their cell phones. Implicit and explicit policies also influence the level of enforcement of laws. For instance, though the undocumented are "illegal" in both Maricopa County of Arizona (which includes the city of Phoenix within its jurisdiction) and Los Angeles County, the Maricopa County Sheriff's Department, under the "leadership" of Joe Arpaio, routinely stops "foreign-looking" types to check their papers while the Los Angeles County Sheriff's Department does not.

Foucault does, however, hedge his bets as he acknowledges some limitations to the state and its all-seeing powers. Foucault's determination of the range and extent of power is never centralized around one source: there is no universal gaze nor a cyclopean eye of a monolithic Panopticon. Rather, power is diffused here, there, and everywhere.

Conclusion

Foucault's conception of the milieu and Goffman's use of the situation may be combined or superimposed upon one another to craft an instrument to comprehend actions, occurrences, and events. Goffman provides tools by which to batten down the logistics of interpersonal behavior, while Foucault provides tools by which to demarcate the relative homeostasis of that which is occurring: that is, the limits of security vis-à-vis risk and the limits of risk vis-à-vis security. Put them together, and one may have a fairly comprehensive method by which to analyze the immediate actions of the everyday, especially in terms of their sociopolitical ramifications.

The second take-away point is that Hacking is mistaken in inserting Foucault into an exclusively top-down slot. Foucault, much like the everyday, is too elusive to be so easily confined into any one position. And his work is so far-ranging, his corpus so vast, and his mind so overflowing that any attempt to corral Foucault into a pithy analysis is doomed. Perhaps in his writing Foucault wanted to replicate what he observed about the workings of the relations of power. The dissemination of his ideas and the popularity of his writings and lectures bear a certain resemblance to Newton's third law of motion, so that for every instance of a statement of Foucault's that seems to reflect a top-down hierarchical and rigidified view of power we can find a match reflecting a bottom-up cellular and capillary view of power. Another way to look at this is provided by a "medieval saying aimed at capturing the links between alchemy and astronomy," which is referenced by Peter Galison in his *Einstein's Clocks, Poincaré's Maps: Empires of Time*: "In looking down, we see up; in looking up, we see down" (2003, 268).

However, even given whatever caveats one has about Hacking's attempt to align Goffman with Foucault, there is still something intriguing about the suggestion of this combination of the two. And, given that, I devote the rest of this study to filling in the in-between space between Foucault and Goffman and whatever top-down and bottom-up positionality they may assume.

CHAPTER 2

The SpaceTimePlace "Thing"

Prior to beginning the chapter proper, let me outline what I am covering. First is the consideration of time, an analysis of which is provided via an exegesis of Torsten Hägerstrand's time-geography. Then space is added to time before bringing in place, ending with the postulation that space, time, and place are indivisible. While it may be analytically convenient and rhetorically conventional to regard them as distinct, such a separation is essentially delusional and, hence, categorically unscientific as well. However, the main purpose of the chapter is to bail out "situation" and "milieu" from their deterritorialized, despatialized, and atemporal *situation* and *milieu* by giving them the aid of time, space, and place, or, as I argue, by the insertion of space-time-place into my schema. As the everyday cannot exist without space, time, and place, this is a necessary operation, one that should yield results in the quest for the elusive everyday.

Hägerstrand and Time-Geography

"Of equal importance is the fact that time does not admit escape for the individual" (Hägerstrand 1970, 10). I begin with this citation of the Swedish geographer Torsten Hägerstrand for two reasons. First, of course, its focus is a large part of the focus of the present chapter, that is, time and its impact on the everyday. Second, this citation matches up with Blanchot's epigram: "Whatever its other aspects, the everyday has this essential trait. It escapes" (Blanchot 1987, 14). What I intend to underline here is the notion that what we must concentrate on is that from which we cannot escape, whether it be time, the situation, place, bodily idiom and its continuous expression, space, the everyday, and so on; and that this notion is counterbalanced on some sort of metaphysical fulcrum with that which we cannot pin down, that is, that which is always escaping, whether that be time, the situation, the milieu, place, bodily idiom and its continuous expression, space, the everyday, and so on. In other words, what we cannot escape is also always escaping. This bind seems to be a reciprocal operation: that which we cannot escape and that which is always escaping locked together in a mysterious embrace, centripetal and centrifugal forces unable to extricate themselves from one another.

Let me first attempt to delineate the basic outline as well as define the components of Hägerstrand's time-geography prior to making more general remarks about time itself and how it fits within the structural apparatus that is being formulated here.

"Continuity and corporeality," states Hägerstrand, "set limits on how and at what pace one situation can evolve into a following in a purely physical sense" (1982, 323). At least to a certain degree, this distills time-geography down to one succinct statement. Fortuitously, it can also serve as a marker in this investigation, as it contains references to the terms essential to this conjuncture of our examination: time, space, situation, and place, though the last may be a bit attenuated here. The pace, or the velocity, at which one situation can evolve into another, of course, references time. Space is included by the references to continuity, corporeality, and the "purely physical sense." This alludes to the physical limitations imposed upon people qua human beings. We must move with our bodies in tow, spaces are available to us only via the limited mobility as well as the required spatiality of our corporeality, and spaces must be serially traversed through or across contiguous places. Even if I fly in a supersonic jet from, say, San Francisco to Tokyo, I must move my body into that jet through a series of movements across contiguous spaces and remain in the place of the airplane's cabin during the flight; then, once disembarked from the jet, more spaces must be traversed via contiguous places before I arrive at my final destination. Much as I may want to be transported in one fell swoop from Point A to Point Z without having to move through any intervening points, I cannot (as of yet) transport my body anywhere without physical limitations "interrupting" my mobility. As Hägerstrand puts it, "Jumps of non-existence are not permitted" (1970, 10). Hägerstrand terms the routes through which and by which we move "paths" and the motivations that kick-start those paths into action "projects;" in turn, paths and projects occur against settings or in places. Hägerstrand refers to this last item as a "diorama."

"People are not paths, but they cannot avoid drawing them in space-time," Hägerstrand states in "Diorama, Path, and Project," probably his most celebrated statement of the basic principles of time-geography (1982, 324). Stating that people cannot avoid drawing paths through space-time may seem so obvious as to be not worth stating. According to Alan Pred: "Hägerstrand is claiming, in other words, that existing scientific principles simply cannot get at certain process outcomes because they ignore the 'collateral' nature of processes involving the actions and experiences of 'individuals,' and because they fail to recognize that all such processes unfold in unbroken temporal and spatial continua" (1977, 211). So, according to Pred's evaluation of time-geography, what science "cannot get at" are the "*local connectedness*" of everyday actions as they are expressed spatially and temporally (Pred 1977, 211; italics, Pred).

"What I have in mind," states Hägerstrand, "is the introduction of a time-space concept which could help us to develop a kind of socio-economic web model" (1970, 10). By tracing spatial paths through and between places and then inserting the element of time, Hägerstrand hopes to complement abstract socioeconomic conceptions with "actual" spatial and temporal components. This can only be accomplished through a concentration on what Pred refers to as "real-world phenomena," that is, actual things happening to actual people in actual places (Pred 1977, 211). By doing this, general "constraints . . . and abstract rules of behavior" can be given "a 'physical' shape in terms of location in space, areal extension, and duration in time" (Hägerstrand 1970, 11).

Let me pause here to plug the components of Hägerstrand's schema into the holes of Goffman's analysis. What is missing from Goffman (or so I assert) is history and geography, time and space. What Hägerstrand is proposing is "a time-space concept" that can be developed into "a kind of socio-economic model" prior to being applied to "real-life phenomena" as they are manifested in paths, projects, and dioramas. Thus, it would seem (that is, if Hägerstrand's warrant can be certified) that Hägerstrand's schema could be nicely aligned with Goffman's. Or at least that's a possibility. For, as Giddens claims in *The Constitution of Society*, with the analytical and methodological tools provided by time-geography, "We are able to flesh out the time-space structuring of the settings of interaction which, however important Goffman's writings may be, tend to appear in those writings as given *milieus* of social life" (1984, 116). In other words, they are given as given and not analyzed at all. Foucault's construal of the milieu, with its limits of security and risk, can also be aided by a more rigorous application of time and space, though this is a more limited claim, as he does, of course, insert the milieu into a variety of geographical and historical frames.

Hägerstrand imposes three sets of constraints on his model: those of capability, coupling, and authority. Capability constraints refer to the limitations imposed on "the activities of the individual because of his biological construction and/or the tools he can command" (Hägerstrand 1970, 12). Notice here that the somatic has been brought into the schema. The constraint of the body, the imposition of indexical positionality, the limitations of the possibilities of physical accessibility, even with "the tools *he* can command," serves as a constraint upon the body as it is located in space and time. "Some" of the capability constraints "have a predominant time orientation, and two circumstances are of overwhelming importance in this connection: the necessity of sleeping a minimum number of hours at regular intervals, and the necessity of eating, also with a rather high degree of regularity" (Hägerstrand 1970, 12). (Adumbrated here is the everyday element of reproduction, an element revisited in the discussion of Marx's notion of reproduction in chapter 4). "Both needs determine the bounds of other activities as continuous operations" (Hägerstrand 1970, 12). Consider that claim from a scientific perspective: sleeping and eating "determine the bounds," both spatially and temporally, of the performance of the continual operations of activities other than those of sleeping and eating. In other words, at some point in time and at some location in space, we must desist from whatever we are doing in order to sleep and eat. Now this may seem so clear that its overt enunciation may seem unnecessary, or even ludicrous, yet the practitioners of the "rigorous" methodology of location analysis neglect precisely these factors.

As a brief sidebar, the foregoing is a perfect example of the neglect of the given: that which is so ingrained into the fabric of the everyday that it is not even noticed. How can any spatial analysis of human activity ignore such crucial components as the constraints imposed by the necessity of eating and sleeping? Because they are so visible they are invisible: "The distinctiveness of the everyday lies in its lack of distinction" (Felski 2000, 17). And that which *completely* lacks distinction *completely* merges with its background—which, of course, means that it *is* the background, and it is treated as such: that is, it is not treated at all!

"Other [capability] constraints," Hägerstrand continues, "are predominantly distance oriented, and as a consequence, enable the time-space surrounding of an individual to be divided up into a series of 'concentric' tubes or rings of accessibility" (1970, 12). In other words, we can only go so far before we must return. "Assume that each person needs a regular minimum number of hours a day for sleep and for attending to business at his home base," whether that base be an actual home, a hotel room, a friend's couch, or a ditch on the side of the road (Hägerstrand 1970, 12). "Attending to business" can be read as a euphemism for the maintenance of grooming norms and the necessities of embodiment. "Everyone, from the most famous to the most humble, eats, sleeps, yawns, defecates; no one escapes the reach of the quotidian" (Felski 2000, 16). "When he moves away" from home base, Hägerstrand continues, "there exists a definite boundary line beyond which he cannot go if he has to return before a deadline" (Hägerstrand 1970, 12–13). In this construal, the spatial ("a definite boundary line") and the temporal (the time required "to return before a deadline") are invoked to denote the limits of one's "concentric tubes or rings of accessibility." "In his daily life everybody has to exist spatially on an island. Of course, the actual size of the island depends on the available means of transportation, but this does not alter the principle" (Hägerstrand 1970, 13). Perhaps this can also be read as an attempt to bring the friction of distance into the realm of the quotidian.

Coupling constraints, the second set of constraining limitations of Hägerstrand's schema, refer to the various locations, times, and durations by which "the individual has to join other individuals, tools, and materials in order to produce, consume, and transact," as the individual fulfills the terms of the project in which she is engaged (Hägerstrand 1970, 14). Since many activities require individuals to form conjunctions in time and space as they use tools and materials to complete projects, such constraints become necessary. We cannot work together at a location unless we are at that location on the same date and at the same time. Also, if workers show up at the same date and the same time but at places that are not coordinated (i.e., places that are different), they will not be working at the same location and therefore will not be working together. This stipulation may have to be modified somewhat, given the spatial and temporal flexibility made possible by working on computers and other modes of modern technology. Still, even given the adjustments necessary to bring Hägerstrand's schema into alignment with the so-called Information Age, if projects are not geared into time and coordinated with space, they will not occur. Deadlines, schedules, and worksites remain, no matter how "virtual" we may imagine ourselves to be. This is true even if these strictures are self-imposed and occur at home. But Hägerstrand is analyzing time and space at the latter stages of what we now call the Fordist era, and his conceptions are perhaps more in alignment with that era than our own.

"We may refer to a grouping of several paths as a 'bundle'" (Hägerstrand 1970, 14). People's paths converge to form "activity bundles" as they work on projects in dioramas. "In the factory, men, machines, and materials form bundles by which components are connected and disconnected. In the office, similar bundles connect and disconnect information and channel messages. In the shop, salesmen and the customer form a bundle to transfer articles and in the classroom, students and teachers form a bundle to transfer information and ideas" (Hägerstrand 1970, 14–15). Though this may seem routine to the point of utter banality, its diction more aligned with the discourse of a simplistic primer than a sophisticated analysis, Hägerstrand recognizes that not all projects are constituted in such a mundane form. Though many activities do depend on near-absolute spatial and temporal coordinations—such as the situation of a factory or a classroom—many operate on a more flexible basis. "Shops, banks, doctors and barbers permit random access between given hours," though here, of course, the datedness of this material betrays itself, as randomly accessing a doctor or even a barber has become a quaint relic of a bygone era, and online shopping possesses a tangential as well as a tendentious relationship with both time and place (Hägerstrand 1970, 15). Yet and still, the general principle holds.

The last constraint that Hägerstrand applies to time-geography is that of authority. This form of constraint refers to "domains . . . the insides of which are either not accessible at all and are accessible only upon invitation or after some kind of payment, ceremony, or fight" (Hägerstrand 1970, 16). Here we can imagine many a setting, from a professional basketball game to a conference at M15 headquarters to a frat party, all domains "within which things are under the control of a given individual or a given group" (Hägerstrand 1970, 16). It does seem that there may be an overlap with coupling constraints here, as a factory floor, one of the sites "coupled" with the concept of coupling constraints, would also fall within the domains falling within the constraint of authority as well. But to do so would be committing a category mistake, as the constraint of coupling is oriented more toward projects the completion of which requires that linkages be made with other people, tools, or materials, while the constraint of authority is oriented more toward places access to which is controlled by stipulated individuals or groups.

The outline of authority constraints given by Hägerstrand moves within close proximity of Goffman's notions regarding possessional territory and the claiming of stalls. Hägerstrand states that "some smaller domains," which come under the spatial purview of an authority constraint, "are protected only through immediate power or custom, e.g., a favorite chair, a sand cave on the beach, or a place in a queue" (1970, 16). In fact, an entire section in Goffman's *Relations in Public* deals specifically with queues—"*The Turn*" in the chapter titled "The Territories of the Self." In this section, Goffman incorporates the notion of a "negative" queue, "namely, an ordering of persons who are to receive something they do not want, such as a place in a gas chamber. . . . Naturally, here one would be allowed to take any turn ahead of one's position but disallowed from stepping behind or giving up entirely one's position" (1971, 36n11).

Hägerstrand contrasts the constraints subtending smaller domains ("a favorite chair, a sand cave on the beach, or a place in a queue") with authority constraints "of varying size," having "a very strong legal status: the home, land property [sic], the premises of a firm or institute, the township, state, and nation" (1970, 16). Of course, some of these constraints have their own constraints. "A very strong legal status" is not equivalent to an inviolable one: for example, if one fails to pay property taxes, the authority constraining access to a home may collapse; if a property is not properly maintained, it may be declared a nuisance and its ownership nullified; and so on. Other constraints, according to Hägerstrand, "are only temporary such as a seat in the theater or a telephone booth at the roadside" (1970, 16). Once again, we have been brought into propinquity with Goffman, as this falls into alignment with the latter's construal of a stall: "The well-bounded space to which individuals can lay temporary claim, possession being on an all-or-none basis" (Goffman 1971, 32). In fact, Goffman also uses the examples of a theater seat and a telephone booth to illustrate the concept.[1]

"At this point, I find it difficult to proceed any further without support of a real-world diorama with non-invented situations, paths and projects," states Hägerstrand toward the beginning of his essay "Diorama, Path and Project" (1982, 326). This statement reflects our own situation, and so, following Hägerstrand's off-stage cue, we turn to his exemplar of "a real-world diorama" in order "to proceed any further" (1982, 326). Hägerstrand makes the choice to "try to call forth a diorama of which I was once an insider during my first formative years," which happens to have occurred in a small village in "the inner woodlands of southern Sweden" during the first two decades of the twentieth century (1982, 326–27). "The *locality* was a valley with wooded slopes. At the bottom of it ran three parallel lines of communication. In the middle a small river took a breath on its way to the Baltic between little waterfalls every second kilometer. On one side twisted a road, not a major one. On the other side the trunk line between Stockholm and Malmo cut off the woods with a sharp edge" (Hägerstrand 1982, 327; italics, Hägerstrand).

It is perhaps a bit unfortunate (at least in terms of the long-term viability of time-geography itself) that the details of Hägerstrand's autobiography seem more attuned to a nineteenth-century Scandinavian fairy tale than to a twentieth-century scientific hypothesis, but there they are nevertheless, seemingly dated before their exposure to the light of day. However, this choice is meant to be purely contingent; that is, the ideas Hägerstrand is attempting to articulate are not intended to be particular to this small hamlet in southern Sweden but are meant to be universal, conforming to what he later called "a neutral system of concepts with the capacity to mediate between different worlds of thought without questioning the applicability of existing knowledge" (2004, 322). Or, as Solveig Martensson puts it in his *On the Formation of Biographies in Space-Time Environments*: "Although the three case-studies [examined in Martensson's study] do deal with conditions in Sweden, my primary aim is not to procure and disseminate information about the state of affairs in this country. The studies should be taken instead as examples of general ideas about the relation between the action space of individuals and the structure of society in terms of its rules and routines in space and time" (1979, 30).

And there is a nice alignment between the details of Hägerstrand's village in which he spent his "first formative years" and the details required for a suitable illustration of time-geography, as if the theory manifested itself inside the mind of Hägerstrand as a simple reflection of the Swedish village. But perhaps the suitability is too suitable and the picture too picaresque. For Hägerstrand's sake and for the longevity of his theory, one wishes that he had grown up in industrial Manchester rather than in this throwback to a bygone era.

This town was scissioned, according to Hägerstrand, into three parts. On one end of town was a foundry and machine shop (the *Bruk*); "in the 1920s the chief products were locomobiles [*sic*], boilers and threshers" (Hägerstrand 1982, 327). This small factory foreshadows the scale of a modern industrial scene, pulling it out of strict quaintness, as a settlement of about 150 workers and their families lived there. "About two and a half kilometers to the south-west, a farmers' village spread out on a low drumlin," with a settlement of about 150 living in this area. A school was situated halfway between these two outposts, with Hägerstrand's father as the resident teacher and the Hägerstrand family living upstairs from the classroom, thus giving young Torsten a kind of ideal central vantage point from which to observe the two sides into which this particular part of the world was split, the farm to the southwest as a vestige of the old, the factory to the northeast as a pre-figurement of the new, the school in between, the woods all around, the river winding through.

Diorama in place, Hägerstrand applies his tools of analysis, instruments meant to depict the movement of people through various everyday situations as their projects are manifested in space and time. "On most days the effective size of an individual's" range of movement "is much smaller than the potential size which is delineated by his ability to move."

> The purposes of movement from the home base include going to work, collecting goods, meeting other people, etc. If we look closer at the time-space volume within reach, it turns out to be . . . a prism. It not only has a geographical boundary; it has time-space walls on all sides. Depending on where the stops are located and how long they last, the walls of the prism might change from day to day. However, it is impossible for the individual to appear outside the walls. Every stay at some station means that the remaining prism is shrinking in a certain proportion to the length of the stay. (Hägerstrand 1970, 13–14)

Plotted in congruence with the time(s) and space(s) of his native village, Hägerstrand's time-geography can supply us with a certain graphic representation of everyday life as people move through time and across space. And it can lend a certain scientific basis to daily routines; at the least, time-geography brings daily routines into focus as an object of spatial and temporal analysis. Yet it has not seemed to produce the results it promised, the fate of many a conception proclaiming paradigmatic status for itself.

Perhaps this failure is due to Hägerstrand and the early practitioners of time-geography making too many grandiose claims about both its efficacy and the potential effects of that efficacy. Perhaps a certain hubris doomed the theory from the moment Hägerstrand stated, "A time-space web model . . . should, in principle, be applicable to all aspects of biology, from plants to animals to men" (1970, 20). Or perhaps that hubris was catalyzed when Pred proclaimed that "once one is deeply imbued with the time-geographic mode of thinking a whole new world of insights is apt to open itself up," including insights leading to new and improved analyses of "the intellectual history of an entire discipline, an academic school of thought, or a school of artistic or musical creation"; the reinterpretation of "certain large-scale, historical-developments . . . in the context of knowledge about small-scale, time-geographic realities"; "the phenomenon of alienation, so widespread in modern urban-industrial societies and so complex in its origins"; and "fresh insights into the changing role and form of the family in Western societies from before the 'Industrial Revolution' to the present . . . how changes in the activity system affect the daily path and life path of individual family members, and how, in turn, the individual taking up new activity-system roles and tasks constrains the choreography of existence and other family members" (1977, 217–18). Even with the inserting of certain qualifiers such as "perhaps" and "possible," Pred's claims ring hollow, as their scope and range promise such a bounty that the inevitable failure to fulfill such promise leads to disappointment.

And lead to disappointment it did. Or at least that appears to be the case. Feminists, such as Karen Davies, have complained that "the social relations in which we are embedded," particularly the gendered aspects of those relations, have not been properly reflected in the theoretical model of time-geography. Furthermore, Davies complains, though time-geography certainly "suggests the potential of dealing with individuals' activities in space and time so that complex interconnections become visible—and this indeed was Hägerstrand's vision, the model continues to build on a somewhat limited and traditional understanding of time and space and this pitfall mars its possibility of making sense of the gendered nature of individuals' lives in a satisfactory manner" (2001, 135, 134). However, in "New Landscapes of Urban Property Management," Jennifer Wolch and Geoffrey DeVerteuil use time-geography as a method by which to understand "how the everyday time-space negotiations of marginalized and non-marginalized populations are both products of and influence upon . . . [the] larger processes" involved in "new" forms of "urban poverty management" (2001, 151, 150). Moreover, Risa Palm and Allan Pred use time-geography as their primary method for their working paper "A Time-Geographic Perspective on Problems of Inequality for Women," in which they assert that "once Hägerstrand's 'time-geography' scheme is grasped, the perceptive reader will realize that most of the inequitable activity-choice options confronted daily by women of various marital and age classes need not be placed into separate categories such as job or educational discrimination, but instead can be placed within a general framework that relates to the 'quality' of their entire pattern of existence" (1974, 14).

In her *Feminism and Geography*, Gillian Rose critiques time-geography at a much more fundamental level than does Davies: "Sexual attacks warn women every day that their bodies are not meant to be in certain spaces, and racist and homophobic violence delimits the spaces of black, lesbian and gay communities. . . . Time-geography speaks the feeling of spatial freedom which only white heterosexual men usually enjoy" (Rose 1993, 34). This rings true, at least to a certain degree, for despite their recognition of a range of constraints subtending a variety of spaces, the early practitioners of time-geography seem to assume that space is open, available, affable, suffused with a kind of Newtonian universal easy accessibility. But given the lack of access experienced by so many people to so many places, inaccessibility rather than its counterpart may be the more ubiquitous standard by which to judge the opportunity of smooth and easy ingress into spaces.

Marxist geographers, most notably David Harvey, advance the complaint that time-geography "tells us nothing about how 'stations' and 'domains' are produced, or why the 'friction of distance' varies in the way it palpably does. It also leaves aside the question of how and why certain social projects and their characteristic 'coupling constraints' become hegemonic . . . and it makes no attempt to understand why certain social relations dominate others, or how meaning gets assigned to places, spaces, history, and time" (1989, 211–12). But perhaps Harvey has made a category mistake in his analysis, confusing an analytical tool with a theoretical discourse. Agreed, time-geography does none of the things Harvey thinks it should; however, it could be used as one tool among many to prove or demonstrate exactly such things. In other words, time-geography is a hammer, not a manifesto on the necessity of the use of hammers. And as a hammer, it could be used by Marxists (or fascists or neo-imperialists or flat-earthers, for that matter) for any number of purposes. But the category mistake could be laid at the feet of time-geography's advocates as well, as they seem to have also conflated their tool with a discourse.

However, it doesn't finally matter what one thinks of time-geography or even how one conceives of time, its possible dimensions, or its probable aspects, just as long as one does conceive of it and allow its possible permutations to merge into the system under formulation. And so a brief exegesis of time itself must ensue, with Fredric Jameson and Doreen Massey as guides. This choice is somewhat arbitrary as there are numerous thinkers who have tried to puzzle out time in all its manifestations and ramifications; for the purposes at hand, however, Jameson and Massey offer enough of a foothold on time to serve as at least a good starting point into this knotty subject.

What Is Time?

"What is time?" Fredric Jameson asks. "A secret—insubstantial and omnipotent. A prerequisite of the eternal world, a motion intermingled and fused with bodies existing and moving in space" (2003, 695). As "a prerequisite of the eternal world," it must be included in any schema that hopes to capture the everyday, that is, if "this" world is part and parcel of the "eternal" world, a question that metaphysicians and theologians have struggled with over the centuries. But what definition of time is needed in order to properly batten down a system such as this? Isn't it a bit facetious to argue, as I have just done, that it matters not how we conceive of time, as long as it is *there*, when such a loose-limbed formulation of time leaves one with a formless and therefore timeless and therefore *useless* conception of time? But if time is a secret, as Jameson claims, at once insubstantial and omnipotent, both nowhere and everywhere, then what can one hope to make of it? Won't its elusiveness simply add to the elusiveness already attending the everyday, transporting the present schema into a heap of insubstantiability? Allowing those questions to hang suspended, let me return to the question at hand, that is, the question of time.

"Over and over again, time is defined by such things as change, movement, history, dynamism" (Massey 1992, 72). Such a conception is typically aligned with a construal of time as entailed with progress, a primary aspect of the modern: "Time was the dominant of the modern" (Jameson 2003, 696). However, with the fall of the paradigm of the modern and the subsequent rise of the postmodern, space replaced time as *the* paradigmatic element. Time's *time* was over, or so it seemed. It was the end of history and the dawn of a timeless era in which a conception of globalized space would reign. And thus was enunciated the spatial turn, an attempt at a paradigm shift that retains the concepts of the temporal along with the predominance of the spatial as the reductive binaries that I believe they are not.

"Modernism . . . ended some time ago and with it, presumably, time itself . . . it was widely rumored that space was supposed to replace time in the general ontological scheme of things" (Jameson 2003, 695). Both Jameson and Massey push back against such a simplistic reduction. At a rather fundamental level, Massey argues that "if spatial organization makes a difference to how society works and how it changes, then far from being the realm of stasis, space and the spatial are also implicated . . . in the production of history" (1992, 70). Here, to push Massey's argument a bit further, the notion is that if space is paradigmatic and predominant, then it must be implicated, and implicated deeply, in change and process in general and in historical and political transformations in particular, all of which cannot be formulated or even imagined without the incorporation of some sort of conception of time. Jameson goes even further, pushing time back to a position of superiority: "Always and everywhere we have rather to do with something that happens to time; or perhaps, as space is mute and time loquacious, we are able to make an approach to spatiality only by way of what it does to time" (2003, 706).

Jameson's notion of an interplay between two distinct entities, time and space, with space mute and time loquacious, creates a pivot point upon which the present argument can turn, as I want to put forward the contention, which I assume to be extremely noncontroversial, that there is no time without space or no space without time, and, furthermore, that the continual predilection to assume that there are leads to theoretical incoherence. For while theorists of sundry stripes may pay lip service to "spacetime" or "time-space" and may also quickly acknowledge that Einstein's theories have made the separation of time from space a notion tinctured with a quant sense of Newtonian antiquity, space and time are still theorized as if they can be separated and held up to the light as illuminated and luminous autonomous entities.

But before tackling that subject, let me return to time-geography to garner what is viable while casting aside what should be discarded. First, it seems to me that the two-dimensional rendering typical of time-geography's modeling methodology does the theory a fatal disservice, as flat replications of the four dimensions of time and space rob time-geography of both space and time. Perhaps if time-geography had been formulated with the prosthetic aid of a program such as Auto-Cad, it would have been more cogent as well as more perdurable. As it is, the provenance of Hägerstrand's theory in the diorama of early twentieth-century village life in Sweden combined with its representation in two dimensions keeps the theory tied to the two pillars of the past and the page. I am not certain, of course, that any sort of union with computer technology would have salvaged time-geography (and perhaps there are geographers working on such a model right now), as no conclusions can be drawn from counterfactuals, but I do think that if such a combinatory methodology had been attempted, it would have had a better chance to gain a purchase on what time geographers were trying to theorize and represent.

In "Matter(s) of Interest: Artefacts, Spacing and Timing," published in 2007, Tim Schwanen proposes an "alliance" between time-geography and Bruno Latour's Actor-Network-Theory (ANT) as just the thing to rescue this brand of "'older' geographical work . . . which might [still] be useful to recent material geographies" (2007, 9). In particular, since both the schemas of ANT and time-geography "are process-oriented and performative, highlighting actual practices, materiality and issues of transportation and transformation," Schwanen believes their frameworks might prove to be compatible and might "benefit ANT and "(post) ANT" theorists in the consideration of "how objects come to hang together locally with other entities present and absent in the landscape and how local connections bear on the durability and reach of sociomaterial assemblages" (2007, 14, 19). The coupling of geographic information system (GIS) technology and time-geography has been attempted—"GIS provides an effective environment for implementing time-geographic constructs and for the future development of operational methods in time-geographic research"—but doesn't seem to have led to any sort of promised land in which GIS-inflected time-geography has come to predominance (Kwan 2004, 268). Whether any of these combinations would indeed be beneficial goes beyond the scope of both this enterprise and my expertise; I merely note the suggestions to indicate the possibility of connecting time-geography to various other schemas and techniques.

In *Innovation Diffusion as a Spatial Process*, Hägerstrand himself predicts the end of the utility of time-geography, given certain conditions: "In a society where there are no appreciable time or cost obstacles preventing one individual from coming into contact with any other individual, relations within 'social space' cannot be appreciably modified by the constraints of geometrical space" (1967, 7). Yet even though we have come much closer to such conditions, what with the profusion of such technologies as the computer and the cell phone along with such applications as YouTube and Facebook, we do not seem that much closer to being a one-point society in which geometric space has become nugatory. People are still living in distinct places, no matter how "connected" they may be, not to mention the fact that billions are not hooked into the Information Super Highway in any way, shape, or form. If time-geography has not gained a purchase on the general intellectual imagination, or if it has not even become the primary methodology within the discipline of geography, this is not due to the transformation of separated "points" into one point.

Still, time-geography did manage to provide the service of at least making the attempt to orient space with time and time with space. It brought everyday routines into the ambit of geographical investigation. Stores, banks, workshops, and factories, coordinated with the time spent in such locations, became sites of investigation as the ordinary and the routine became foci of examination. Whether the practitioners of time-geography did this in the most complete or most cogent way is certainly debatable, but at least they made the attempt while others were doing geography as if space and place were outside of time, reifying them as scientific objects aloof from the friction of time and the necessities of space and place.

And so time is brought into my schema. But what time? Time as what? Am I gesturing toward a progressive or a cyclical time? An incremental time or a time expansive beyond a measurement in degrees? Time as "(1) sacred, (2) profane, (3) micro scale, (4) synchronized, (5) personal, (6) biological, (7) physical" or time as "(8) metaphysical" (Crang 2005, 206)? A time bound to the eternal or a "temporary" time? Should I be scuttling to philosophers of time such as Bergson, Sartre, or Heidegger, the last of whom posits that "temporality temporalizes itself as a future that makes present, in the process of having-been" (1996, 321, § 350), a knotty formulation the untangling of which may or may not clear the thicket that lies ahead? Sequential time? Time as *timing*? Rocky Mountain Time? Greenwich Mean Time? Physical time (whatever that is)? "*Mundane* or *Typological Time*" (Fabian 1983, 30; italics, Fabian)? "Intersubjective time" (Fabian 1983, 1)? Biographical time? *Auto*biographical time? Diachronic time? Synchronic time? Let us leave these and a thousand other prickly temporal questions aside, at least for *now*, and conclude this section with one irrefrangible postulate: time cannot be conceived without space.

Add Space

What I want to do in this section is supplement time-geography's merger of time and space with some elementary physics, buttress it with some theoretical work from geography and assorted other disciplines, and then reinsert my findings about time and space into my schema.

"At one level of analysis," Pred indicates, "time-geography deals with the time-space 'choreography' of the individual's existence at daily, yearly, or lifetime (biographical) scales of observation. Time and space are seen as inseparable. Each and every one of the actions and events which in sequence compose the individual's existence has *both* temporal and spatial attributes—not merely one or the other" (1977, 208; italics, Pred). Could it be possible that if space and time are *seen* as inseparable, they *are* inseparable? And if this is granted, then could it be possible, or even plausible, that they are one and the same? Here, I am not merely postulating that any definition of "space is neither self-evident nor self-sufficient but is rather often mutually and problematically defined by and with problematic concepts of time" (Crang 2005, 200). Instead, I am claiming that time and space are so entwined that thinking of them as distinct entities is a category mistake of the most basic kind. As Doreen Massey states, "Space must be conceptualized integrally with time," as indeed it would have to be if space and time are inseparable; "indeed, the aim should be to think always in terms of space-time" (Massey 1994, 2).

What I want to suggest is that geographers and others take the term "spacetime" literally, as if its conjunction truly equates with its status as a unified entity. As Crang himself concludes by the end of his essay on time and space, the "inseparability" of time and space "is not just a matter of bolting two conceptually discrete elements together, but rather that the two are not separable conceptually" (2005, 217).

This is a step further than Edward W. Soja pushes the inseparability of the two, as Soja writes that "*we are just as much spatial as temporal beings . . .* our existential spatiality and temporality are essentially or ontologically coequal, equivalent in explanatory power and behavioral significance, interwoven in a mutually formative relation. Human life is in every sense spatio-temporal, geo-historical, without time or space, history or geography, being inherently privileged on its own" (2010, 16; italics, Soja). So, though Soja is willing to hyphenate our two protagonists, he doesn't seem to be quite willing to take the final leap and tip their putative equivalence into absolute unification, perhaps because it would mean the death of the spatial turn, as spatiality would no longer be an autonomous unit. Soja's notions of the explanatory equivalence of the spatial and the temporal, along with his strenuous but ultimately misguided effort to boost space into theoretical equality or predominance over time, should be discarded. And so should Crang's qualifier "conceptually," which he appends to the inseparability of time and space. Let us instead assert that time and space are not separable at all, neither conceptually nor actually, period, end of story, finis! And, being identical, they cannot be equivalent. The two are one, time and space a unified whole. If "space cannot be divorced from time," as the philosopher Jeffrey Malpas claims, perhaps this is simply because they are not the types of things that can either be married or divorced, so that even these metaphors in regards to time and space are essentially category mistakes and therefore essentially incoherent (1999, 42).

"Time is not a quantity in itself . . . it cannot be separated from the motion of physical bodies" (Sambursky 1962, 13). Even the devices by which time has been measured cannot be separated from physical bodies as the very physicality of bodies necessarily invokes space. A sundial's dependence on the movement of the sun, a watch's reliance on the slippage of its escapement, an hour glass's combination of its narrow portal with its grains of sand—the functionality of all these devices measures time by or through various modes of motion across or through space. As Leibniz writes in his third letter to Clarke: "Instants, consid'd without the things, are nothing at all . . . they consist only in the successive order of things" (Leibniz and Clarke 1956, 27). Even internal timekeeping, embodied gauging of time, depends on the space of bodies to keep track of the flow of time. But the sheer radicalism of Samuel Sambursky's statement that "time is not a quantity in itself," the metaphysical ramifications inherent in that statement, may not be properly understood without a more incisive analysis of its conceptual core. For if this statement is true, then the presupposition that somewhere, somehow, actual units of time are ticking away is a fantasy, pure and simple. The universe has no clock. We invented time, in all its manifestations.

Such a realization does not mean that change and process do not exist, or that cyclical or even progressive events do not occur in circular or sequential order; merely that quantitative units of a temporal kind only exist in our spatial relationship to them. We invented various methods of calculating time, have found these more or less convenient, and then naturalized some and rejected others. In so doing, we have come to think of time as somehow existing in a sort of nonspatial domain, as if there could be such a thing as a domain without spatiality. Ignoring this colossal mistake, we have continued on as if time marched on as well, in some receptacle without dimension, sans shape, sans size, sans motion, a spaceless ticking hub counting down units in a distinctly zero-sum game.

"All judgments concerning time," writes William Lane Craig of Einstein's special theory of relativity, "are statements about *simultaneous events*" (2001, 28; italics, Craig). Such events can only occur when time is coordinated with events occurring in places. "To use his [Einstein's] example," continues Craig, "when I say 'The train arrives here at 7 o'clock,' I mean, 'The train's arrival and my watch's pointing to 7 are simultaneous events'" (2001, 28). Given this, time only exists as simultaneous events exist, which leads to my next inquiry: what are events, at least as indicated by Einstein's example? One is the arrival of a train at a station, a coordination of the motion of a body, the train, coming to a rest in a place, the station. But the train itself is also its own place, as experienced by those traveling inside it. And the looking at the watch, the noting of the watch as indicating "7," is also construed as an event; this event must also exist in a place situated in space as well.

So time is seen here as so enmeshed with space that it cannot be calibrated without a coordination with bodies moving through space. Here one might want to consider the "temporal" unit of a year. Obviously, a year is just as much a spatial as a temporal unit, for what is a year except the completion (more or less) of the Earth's ellipse about the sun, so that a movement in space (an event) is determining our notation of a calendar year. We want to say, or tend to think, that this is caused by the Earth's orbit and, in turn, if further explanation is needed, that the orbit of the Earth is caused by the gravitational "pull" of the sun, but "gravitation, as understood by the theory of general relativity, is to be comprehended in the geometric structure of space-time" (Jammer 1954, 163–64), a slight detail of physics we tend to ignore as a by-product of our naturalization of time's duration.

Much of our difficulty with our conceptions of space and time (or space-time or time-space, or whatever else we wish to call "it") has to do with language and its tendency to hypostasize that which it names. The creation and use of the two words "time" and "space" has implanted in our minds the notion that there are two distinct things replicating the distinctness of the two terms, a stubborn thickness of thought conflating the name and the named, leading to a confusion between the metaphor of the linguistic with the actuality of the existent. Perhaps it is a case of what Milic Capek calls "sheer semantic inertia, a simple concession to our traditional and outdated linguistic habits" (1961, 190–91). But even given such inertia, as Allan Janik and Stephen Toulmin state in *Wittgenstein's Vienna*: "No axiomatic system can by itself *say* anything about the world. If such a system is to perform a propositional—that is, a linguistic—function, something more is required: it is necessary to demonstrate, in addition, that the relations actually holding between language and the world make such a formulation possible" (1973, 188–89; italics, Janik and Toulmin).

My contention is that "the relations actually holding between language and the world" *do not make possible* the distinct formulation of the terms "space" and "time." They are a vestige of an overturned world; yet, they linger, residue of a former epoch, persistent in their hold upon our conception of the world, their presence habituated into the habits of our psyches and the uses of our everyday language. Of course, I must acknowledge that in my attempt to untie this linguistic knot, I am also attempting to "fix" language through the use of language, a self-referential meta-operation of the hapless and hopeless variety.

"Time and space are not passive frames for action but part of the action itself" (Holt-Jensen 1999, 127). This seemingly simple and almost rustic formulation belies the extreme degree of difficulty of the realization of that which the sentence denotes. Time and space are not "sitting out there," ready-to-wear, passively waiting in some offstage metaphysical vestibule for our conceptual schemas to scoop them up and put them into action: they *are* the action, always and already, all the way down. Not only that, but they are mutually constituted, their "genealogies" so intertwined that to suggest that they have distinct genealogies is not only fallacious metaphysics but sophomoric physics. It isn't only, as John Agnew says, that the separation of space and time "is intellectually and politically untenable" and that "each requires the other to fulfill its potential," but that there are no two distinct things able to harbor separate sets of potential to be fulfilled (2005, 92). Agnew makes the cogent point that the idea that space can be scissioned off from time is politically as well as intellectually untenable, as the abstraction of space and its reification into coordinate systems has led to the loose-limbed logic of capitalism that mistakes actual places for localized (i.e., coordinated) spaces and then exchanges them as if they were eradicated from the soil in which they exist. But the basic sense of his diction still treats the entities, time and space, as if they are separate things.

"Concentrated on looking for the hidden which we think is there, we have no attention to spare for what lies openly around us" (Paul 1956, 92). While we gaze deeper and deeper into space and calculate time down to the nth degree, we tend to ignore the fact that not one iota of space has ever been observed outside of time and that time and space have never been disarticulated, no matter how much their disarticulation assiduously resists articulation (that is, if the disarticulated can be loaned the necessary agency to bequeath it the power to preempt articulation). In *Space, Place, and Gender*, Doreen Massey cites the physicist Hermann Minkowski as stating that "space by itself, and time by itself, are doomed to fade away into mere shadows, and only a kind of union of the two will preserve an independent reality" (Minkowski 1964, 297, qtd. in Massey 1994, 3). And whatever that "independent reality" may or may not be, *it will not contain* independent entities called space and time.

The means of measuring space and time also reflect, to a certain degree, if not the incipient doom Minkowski references, then at least the less-than-precise metaphysics *and* physics that such measurements seem to imply. The metrics used for the measurement of bodies, a spatial calculation, are susceptible to the same labile qualities as the metrics used for the measurement of time, a calculation of bodies in motion. Such an axiom not only connects time and space but interweaves them to such a degree that they become indiscernible. To test this, discern space without time or time without space: impossible! Though conceiving of space and time as a unity may seem "more difficult" as well as more "paradoxical" than maintaining their conceptual distance from one another, "we could say," to quote Einstein and Infeld, "[that] modern physics is simpler than the old physics and seems, therefore, more difficult and more intricate. The simpler our picture of the external world and the more facts it embraces, the stronger it reflects in our minds the harmony of the universe" (1966, 213). Now while one can postpone any final adjudication concerning the harmony of the universe and its (possible) relationship to the present conception of space and time, what one cannot do, or perhaps what one *should not do*, is maintain the pretense of the inviolable distinction of the spatial from the temporal, a pretense practiced in everyday usage as well as in academic discourse. As Michio Kaku states in *Einstein's Cosmos*, what Minkowski's "mathematical language" proves is that "space and time" form "a four-dimensional unity" (2004, 73).

Citing Poincaré, William Lane Craig also probes the question of what it is we are exactly gauging when we measure time and space:

> If all objects, including our measuring instruments, are deformed according to some law, "we will not be able to notice this. . . . In reality space is therefore amorphous, a flaccid form, without rigidity, which is adaptable to everything; it has no properties of its own. To geometrize is to study the properties of our instruments, that is, of solid bodies." The same thing holds for time: if everything went more slowly, we would not notice it. "The properties of time are therefore merely those of our clocks, just as the properties of space are merely those of the measuring instruments." (Craig 2001, 32).

But, of course, this similarity does not lead to the conclusion that space and time are identical or isomorphic; merely that our calculations of them refer to our instruments of measurement and not to what we refer to as space and time. The contingency of these measurements is reflected in the histories of both the adoptions of the calendar year as well as the prime meridian. In the case of the latter, given that the Greenwich Mean Line was not adopted as the "'common zero of longitude'" until 1884 (with the French holding out for Paris until 1911, when they finally acceded to the selection of the Royal Observatory as the Prime Meridian), it is little wonder that zero meridians sprouted up all across the globe (Strong 1935, 481, qtd. in Pratt 1942, 234). Indeed, up to 1870, such meridians ran through numerous locations, including "Greenwich, Paris, Ferro, Naples, Christiania, Copenhagen, Brussels, Madrid, Cadiz, Pulkova, Rome, Stockholm, Lisbon, Amsterdam, Rio de Janeiro, and Washington" (Pratt 1942, 233–34). In 1932 Hendrik Willem Van Loon reported that "even today . . . there still are German, French and American maps which show" the prime meridian "running through Berlin, Paris and Washington" (1932, 64).

All of these are examples of what Bourdieu calls "arbitrary necessity," a "collective belief" in contingent limitations that ignores its own ignorance of the arbitrary quality of that necessity: "This magical act presupposes and produces collective belief, that is, ignorance of its own arbitrariness (Bourdieu 1990a, 210). I cite Bourdieu here to support my thesis of the arbitrariness of the general distinction between space and time and of the particular modes of distinction between them as well.

However, the only thing this proves, or merely supports, is that the measurement of space and time are contingent. It does nothing to support the idea that these spatial and temporal systems of measurement are isometric, nor does it bolster the notion that space and time are identical. So, perhaps one should take a further step and draw the conclusion that the temporal and the spatial are distinct entities. Besides, what does one gain or lose by adopting the cockeyed notion that they are the same? Hasn't one benefited by separating out time from space and space from time? What if the nations of the world were still starting their years on different days (Latouche 1996, 23)? What if prime meridians still ran through Ferro and Copenhagen, Cadiz and Pulkova, let alone Paris, Naples, Christiania, Brussels, Madrid, Rome, Stockholm, Lisbon, Amsterdam, Rio de Janeiro, and Washington? Nothing would be coordinated: spatial and temporal cockamamy would reign supreme. If we reckoned space and time to be the same, we would have no way of differentiating them; then we might end up conflating the time of day with a shape or a shape for the time of day. But isn't that exactly what we do when we point to a shadow cast across a sundial or calculate the angles of a watch? We are conflating shapes and angles with times of day! "A time measurement . . . has meaning only in reference to a reference frame having a specified state of motion," states Einstein in his "Uber das Relativitatsprinzip und die Folgerungen" (Einstein, qtd. in Craig 2001, 25n14).

So I suppose all I am asking is that this conflation be recognized for what it is, a convention performed for the sake of convenience, which does not reflect that which cannot be separated on either a conceptual or a realistic basis, that is, time and space. Though "it seems difficult to conceive and is perhaps unimaginable . . . there is no reference frame 'space' or preferred frame coextensive with space"; indeed, "there is no such frame as space. All that exists is locally moving frames, and even these do not move relative to space, but only relative to each other" (Craig 2001, 38). What trips one up is a habitual adherence to Newtonian notions. Craig cites Milic Capek as confessing that "we are all unconsciously Newtonians even when we profess to be relativists, and the classical idea of world-wide instants, containing simultaneously spatially separated events, still haunts the unconscious even of relativistic physicists; though verbally rejected, it manifests itself, like a Freudian symbol, in a certain conservatism of language" (Capek 1961, 190–91, qtd. in Craig 2001, 38).

Before moving on to a consideration of place and its inclusion in this schema, I want to return to Soja and the spatial turn. First, notice that the term seems to reference time as much as space, for the *turn* is temporal as well as spatial. The spatial turn, according to Soja, occurs in a certain period, roughly toward the end of the twentieth century. It's occurring in time, then, this turn to the spatial, so the spatial turn is every bit as temporal as it is spatial. And if it is becoming less and less possible to separate history and geography, time and space, then why privilege one over the other? The mere switching of the privilege from the temporal to the spatial will simply continue what is essentially a *metaphysical* as well as a *physical* mistake. And if one truly believes in and relies on dialectics, the privileging of the spatial over the temporal will only lead, with the passage of time (oddly enough), to the privileging of the temporal over the spatial. Why not end this seesaw battle altogether and call for a SpatialTemporal Turn?

However, for whatever reasons, Soja doesn't take this route. In his *Seeking Spatial Justice*, published in 2010, Soja states that "as intrinsically spatial beings *from birth*, we are *at all times* engaged and enmeshed in shaping our socialized spatialities, and, *simultaneously*, being enmeshed by them" (2010, 18; italics, mine). This statement, intended (I presume) to highlight the intrinsic spatiality of human beings, is, of course, deeply implicated with temporal terminology. Yet even though Soja states that the "achievement" of "a complementary rebalancing of historical and geographical thinking" requires "at least for the present moment . . . some degree of foregrounding if not a strategic but temporary privileging of the spatial perspective over all others," he doesn't seem to recognize that such a foregrounding, no matter how temporary, will lead, and is bound to lead, precisely to the kind of confusions and inconsistencies reflected in statements such as those cited above (2010, 17).

Here we should also note that Soja deftly raises the stakes, for it is not that the spatial will only be privileged over the temporal, but that it will be privileged "over all others." This is a claim without limits, demanding a privilege for the spatial that will nullify any contender. It is one thing to say, as Massey does, that it "is now [1992] increasingly accepted widely . . . that the social and the spatial are inseparable and that the spatial form of the social has causal effectivity," a perfectly reasonable claim, and quite another to say that the spatial should be privileged over all other comers, a claim that can only lead to an overbearing triumphing of the spatial or to absurdity (1992, 71). This becomes especially overbearing when the claim for the supreme privileging of the spatial is conjoined to a temporal qualifier: such favoritism should proceed only on a *temporary* basis!

"The point here," states Massey, "however, is not to argue for an upgrading of the status of space within the terms of the old dualism . . . but to argue that what must be overcome is the very formulation of space/time in terms of this kind of dichotomy" (1992, 75). Such an overcoming will not be aided by the privileging of either the spatial or the temporal, especially if, as seems to be ceded even by such a spatial thinker as Soja, "space and time are inextricably interwoven" (Massey 1992, 77).

Perhaps I have pushed my argument too far. Perhaps in my attempt to collapse the time/space duality into a singularity, I have simply succeeded in maintaining the distance between the two. But I hope that at the very least I have demonstrated that the implications of relativity have not been fully "taken on board" by geographers as well as by others (Massey 1992, 77). And I also hope that I have clarified the idea that time and space cannot be conceptualized as discrete entities. There is either timespace or there is not. There also doesn't seem to be room to believe that there is some portion of a continuum that is time and another that is space; either it's a continuum or it is not. Otherwise, let's desist from using the term and return to time and space, two divisible entities, divisible absolutely.

However, time and space also cannot be construed or conceptualized without place, the subject of our next section.

Add Place

In *The Fate of Place: A Philosophical History*, Edward S. Casey makes much the same mistake as Soja, but from the opposite direction. Casey is so concerned that "place" claim its rightful position (*place*) in the metaphysical order that he privileges it almost beyond recognition. "It is place that introduces spatial order into the world . . . place provides the primary bridge in the movement from cosmogony to cosmology" (1997, 5). Place, then, is prior to space, for without it, space cannot even be conceptualized, or brought into being, as it were. Referencing various creation myths, Casey states: "In all of these instances, place presents itself not just as a particular dramatis persona, an actor in the cosmic theater, but as the very scene of cosmogenesis, the material or spiritual medium of the eternal or evolving topocosm. Cosmogenesis is topogenesis—throughout and at every step" (Casey 1997, 5).

Let me try to be a bit more modest than Casey about any claims I make about place, but let me first note that Casey doesn't always overshoot the mark. Even in the citation above, he is essentially correct. Space *cannot* be catalyzed into existence, or at least into "spatial order," without the presence of a place in which to do it. As Larsen and Johnson put it in "Toward an Open Sense of Place: Phenomenology, Affinity, and the Question of Being": "Recent phenomenological work points to a sense of place as the singular existential ground for thought, action, and understanding. The insight is that existence is placed: Anything that 'is' first requires a situation to provide both context and horizon for its availability as object. Place is how the world presents itself; that is to say, being inevitably requires a place, a situation, for its disclosure" (2012, 633). Add in their corollary that "the radically empirical observation that any kind of being immediately presupposes its own placing," and one has gone a long way to the justification of place in its metaphysical partnership with time and space (Larsen and Johnson 2012, 636).

But, of course, place cannot do any of this work without the presence of time. Otherwise, *when* would it be done? So, with the addition of place, what

I want to do is suggest that spacetime be discarded in favor of spaceplace-time. Let me cite Hesiod's *Works and Days* for my first authority on such a move:

> When the Pleiades born of Atlas rise before the sun,
> begin the reaping; the ploughing, when they set.
>
> For forty nights and days they are hidden, and again as the year goes round they make their first appearance at the time of iron-sharpening. This is the rule of the land, both for those who live near the sea and for those who live in the winding glens far from the swelling sea, a rich terrain: naked sow and naked drive the oxen, and naked reap, if you want to bring in Demeter's works all in due season, so that you have each crop grow in season. (Hesiod 48)[2]

Here is a nice configuration of space, place, and time formulated as an essentially connected physical and metaphysical concatenation. Movements in celestial space ("When the Pleiades born of Atlas rise before the sun") are coordinated with time ("For forty nights and days they are hidden, and again as the year goes round they make their first appearance at the time of iron-sharpening") and place ("This is the rule of the land, both for those who live near the sea and for those who live in the winding glens far from the swelling sea, a rich terrain"). However, this simple relation or state of being connected one to another, even in the most essential of ways, does not equate with identity or even equivalence. A mother may be essentially connected with her child, but they are neither identical nor equivalent, at least postpartum. However, what this citation does go some way to showing is that definitions and construals of space and time (or spacetime, or time-space) that leave out place are missing an essential item.

Vault beyond Hesiod's era some two thousand years, and we can see that Descartes understood this as well. In the *Regulae*, he states:

> For even though someone may convince himself, if we suppose every object in the universe annihilated, that this would not prevent extension per se existing, his conception would not use any corporeal image, but would be merely a false judgment of the intellect working alone. He will admit this himself if he reflects attentively on this image of extension which he tries to form in his imagination. For he will notice that he does not perceive it in isolation from every subject, and that his imagination of it and his judgment of it are quite different. Consequently, whatever our intellect may believe as to the truth of the matter, these abstract entities are never formed in the imagination in isolation from subjects. (Descartes 1974–86, qtd. in Gaukroger 1995, 170)[3]

In other words, space as emptiness (i.e., "the universe annihilated") cannot be imagined without place (i.e., "extension" and "subject").

If there is not one thing, that is, a unified field of spaceplacetime, then it seems to me there is only one viable alternative, and that would be an infinite variety and number of the same, space and place and time conjoining over and over again, in an endless array of infinite configurations. Massey suggests such a possibility in *For Space*: "if time is to be open then space must be open too. Conceptualizing space as open, multiple and relational, unfinished and always becoming, is a prerequisite for history to be open and thus a prerequisite, too, for the possibility of politics" (2005, 59). Without lingering to comment on every aspect of this passage, there is obviously a leaning here to a conception of space and time (and I would have to insert place as well) as multiple and multiplying.

In a discussion of the ongoing extension of the global codification of universal measurements of time and space, Joseph Rouse makes much the same point but rather more bluntly than Massey: "Not just one space and one time exist, but many, and they are commensurable only to the extent that their commensuration is actively achieved and constantly policed" (1996, 105). Granted, Rouse is discussing the limits of calibration in light of its ever-encroaching imposition upon human consciousness, but he is still making a claim for a multitude of times and spaces, rather than one field uniting place, time, and space. In his *A Foray into the Worlds of Animals and Humans*, the biologist Jacob von Uexküll also makes a case for a multitude of times and spaces. Uexküll argues for discrete "umwelten" for different species of flora and fauna, each umwelt with its own particular cosmos of space and time. And though I concede that a cogent argument can be made for such a multiplicity that stretches asymptotically to infinity, even in such a construal every separate unit consists of a unitary combination of times, spaces, and places that cannot be sundered, even when multiple.

To demonstrate this, let me return to Einstein's example of the watch, the station, and the train. Recall that, as Craig reformulates it, "When I say 'The train arrives here at 7 o'clock,' I mean, 'The train's arrival and my watch's pointing to 7 are simultaneous events.'" Remember that only space and time were under consideration when this example was presented in the previous section. But now notice that the train arrives *here*, the same *here* apparently in which "my watch" shows that it is 7 o'clock. What is this "here" but a place? And even if the watch exists somewhere else, that somewhere else must also be a here, that is, a place. Indeed, how could it be in anything else but a place?

Here, taken as indexical and coordinated with a somatic identifier (a body), having its own particular up, down, back, front, left, and right, must be some place; otherwise it is no place and does not exist.

In *Einstein's Clocks, Poincaré's Maps: Empires of Time*, Galison puts it another way: instead of a total eclipse being conceived as an opportunity to "measure the deflection of the gravitational pull of the sun," Einstein would conceive of it as the "sun's bending of spacetime" (2003, 326). This formulation posits the sun as a place in conjunction with spacetime; I would take it a step further and formulate the triad as one indivisible unit.

Let me take up another example, this from Massey's *Space, Place, and Gender*: "The geography of social relations forces us to recognize our interconnectedness, and underscores the fact that both personal identity and the identity of those envelopes of space-time in which and between which we live and move (and have our 'Being') are constructed precisely through that interconnectedness" (1994, 122). What are "those envelopes of space-time in which and between which we live and move" except places? Indeed, what *could they be* except places? In fact, "envelopes of space-time in which we live and move" goes some way to being a fairly successful and even a rather cogent definition of place. Furthermore, where exactly could we recognize our interconnectedness except in places? And where do we become connected except in places, even if they are places connected through telephonic or Internet links? And if Massey wants to conceive of space-time as what we should always "think in terms of," then place has to be inserted in this conception as well, for where else will we do such thinking? (1994, 2). Even the most spatially oriented thinker must be in a place in order to deny the importance of place!

However, it should be recalled that much of the problem with fitting place into this schema is that "place" as a term lacks a certain coherence or a necessary rigor. Its amorphousness defeats it. With this concession, I am afraid it may be necessary to conduct a brief discourse on the meaning of place.

Place is famously a wobbly concept. On one hand, a car seat can be classified as a place. On the other, so can the world and even the universe. The very plasticity of the concept, its nondiscriminatory catholicity and its rubbery inclusiveness, makes it almost implode into meaninglessness. So perhaps there is a certain justification in the criticism of place as being too soft, too expansive, too fuzzy. In *The Nation-State and Violence*, volume 2 of *A Contemporary Critique of Historical Materialism*, Giddens states, "I use 'locale' in deliberate preference to the notion of 'place' as ordinarily employed by geographers, because 'place' is often only a vaguely formulated notion and because it does not usually mean the co-ordination of time as well as space" (1985, 12). Place is too amorphous for Giddens's purposes, it seems. Giddens also connects "locale" with "setting," which seems to be a necessary architectonic component to his conception of locale. "Locales refer to the settings of interaction, including the physical aspects of setting—their 'architecture'—within which systemic aspects of interaction and social relations are concentrated" (1985, 12–13). The phrase "within which" oddly maneuvers setting, and therefore locale, into conjunction with the term "container," the very word Aristotle uses in his classic construal of place in *Physics*, Book 4: "Since place contains the body, it looks as it might be the *form*; for the extremities of container and contained coincide" (1955, 375; italics, Aristotle). And so it seems that Giddens's deflection of place leads right back to place.

What must be acknowledged by geographers is that "place" is a term that has been left suspended in a kind of definitional limbo. Despite yeoman work by such thinkers as Doreen Massey, Yi-Fu Tuan, Robert Sack, Michael Curry, Jeffrey Malpas, Edward Casey, and Nicholas Entrikin, the conceptual framework of place is still dangling from its hinges, screwed into an apparatus too flimsy to carry its own weight. Some of this amorphousness can be traced to a certain mushiness associated with construals of place. For instance, Tuan, in his *Space and Place: The Perspective of Experience*, asserts that "Place is security" and "Places are centers of felt value" (1977, 3, 4). But what of places that are inherently insecure? Does the resident of a shantytown in Johannesburg make a linkage between place and security? Does the soldier on a battlefield connect the bloody field around him to security? Or the battered wife in her home? A brothel is a place, but is it a "center of felt value"? Only if "value" is taken as exchange value in the form of cash for sex, and only if "felt" references the temporary tactile linkage between the prostitute and the client. In a chapter of *Space and Place* titled "Intimate Experiences of Home," Tuan privileges home as the place par excellence, the primary signifier of place, the prototypical site in which "the sick and the injured can recover under solicitous care. . . . The lasting affection for home is at least partly a result of such intimate and nurturing experiences" (1977, 137–38). But what "lasting affection" for home can an incest survivor have? Perhaps it is this kind of sentimental inscription of place that has exiled the term from usage among many theorists. The vague formulation of place has *displaced* it as the term of choice for that which can only be referred to as place. And so there is "locale," "setting," "space," and "situation" instead, terms that finally can only be understood as stand-ins for place, and terms which also suffer from their own definitional pulpiness.

In *For Space*, Doreen Massey examines the conceptual framework of geographical terminology and finds our understanding of the complexities of place to be sorely lacking. In a chapter aptly titled "the elusiveness of place," she denies that "place" can be associated with a fixed entity, stating instead that places not only shift their identity with the passage of historical time but also shift *places* with the passage of the *longue durée* of geological time. "We use . . . places to situate ourselves, to convince ourselves that there is indeed a grounding. . . . But in the end there is no ground, in the sense of a stable position" (Massey 2005, 131, 137). She ties place to event, arguing for indexical qualities as stabilizing factors for place. "'Here' is where spatial narratives meet up or form configurations, conjunctures of trajectories which have their own temporalities" (Massey 2005, 139). Massey quickly points out that this hooking of temporality to spatiality doubles the indexical quality of place, adding a "now" to a "here." But this may be the price to be paid to begin to approach the complexity of place. In essence, she is arguing for place as understood within the status of the verb rather than the noun. This is the *event* of place in the sense of the coming together of the previously unrelated, a constellation of processes rather than a thing (Massey 2005, 141). But does this complicate place so much that it essentially becomes a cipher? I think not. Instead, Massey's schema of place challenges us to conceive of place in a much more transitive way, as trajectories of the animate and inanimate commingling in indexical equipoise, as a receptacle in which narrative arcs and spatial arcs converge, if only on a temporary basis.

"Another strand of resistance to any affirmation of the importance of place," writes Massey in *Space, Place, and Gender*, comes from "the associations of 'a sense of place' with memory, stasis, and nostalgia: 'Place' in this formulation was necessarily an essentialist concept which held within it the temptation of relapsing into past traditions, of sinking back into (what was interpreted as) the comfort of Being instead of forging ahead with the (assumed progressive) project of Becoming" (1994, 119). Yet while this essentializing aspect of place certainly can be a danger, it can also be one of the strengths of the term. "Place" does connote an affective connection to a site or a location or a given areal territory. And though this connection can descend into stasis and nostalgia, the effect of this affect is not necessarily reactionary. Memory of place can be revolutionary as well, girding people into action as they move to restore the legitimacy of place (whatever scale of place it may be) to its rightful *place*, a rightness kept alive through the affective power of collective and/or individual memory.

There are those who think that, given the problems of "place" as a conceptual term, we should simply discard it and use "territory" instead. However, territory does not catch the affective or memorial aspects of place. Nor is it reflected in everyday language, as it is very difficult to imagine anyone but the most rigidified theorist saying something like, "Isn't this a fantastic territory!" or "Ah, that was a great territory." I am forced to conclude that, even given its admittedly wobbly qualities and its less-than-stellar record as a conceptual *place*-holder, place it is and place it must be, at least until something better comes along.

Now, setting aside that lengthy definitional sidebar, let me return to my effort to combine space, time, and place, which I attempt to accomplish through the use of three examples, taken in reverse order, the first a study of beach erosion in Malibu, the second the "keeping of time" by white dwarf stars, and the last an experiment conducted at Harvard in 1959 that "confirmed Einstein's prediction of gravitational red shift . . . that is, that clocks beat at different rates in a gravitational field" (Kaku 2004, 209).

In the 1959 Harvard University experiment, Robert V. Pound and G. A. Rebka "took radioactive cobalt and shot radiation from the basement of Lyman Laboratory at Harvard to the roof, 74 feet above. Using an extremely fine measuring device . . . they showed that photons lost energy (hence were reduced in frequency) as they made their journey to the top of the laboratory" (Kaku 2004, 209). So here we have a place (the Lyman Lab at Harvard University located in Cambridge, Massachusetts, USA, the Earth), a time (1959, plus the time implicated in the moment in which "they made their journey to the top of the laboratory"), and space (as reflected by both the seventy-four feet from the lab to its roof as well as by the spatial-gravitational-optical implications of the red shift measured in the experiment). If any one of the three components were subtracted, the experiment could not be conducted. Divest the experiment of time, and there is no 1959 nor are there any moments available for the radiation to travel from the lab to the roof. Strip the experiment of space, and there is no distance of seventy-four feet from the lab to its roof nor are there spatial implications for the experiment: no sun in which the red shift occurs and no distance from the sun to the Earth. Take out place and there's no lab, no Harvard, no Cambridge, no Massachusetts, no United States of America, no Earth, no solar system, and, given the most expansive definition of place, no universe.

Kaku adds that in 1977 "astronomer Jesse Greenstein and his colleagues analyzed the beating of time in a dozen white dwarf stars. As expected, they confirmed that time slowed down in a large gravitational field" (2004, 209). Once again, here is a conjunction of time, space, and place, as all three components are irretrievably commingled, whether one considers the lab in which Greenstein and his colleagues conducted the experiment or the gravitational fields of the twelve dwarf stars from which was "analyzed the beating of time" or the time in which it took to conduct the experiment or "the beating of time" within those dozen white dwarves. In fact, the question of the separation of place, time, and space in the operation of "the beating of time" in the gravitational fields of a dozen dwarf stars seems superfluous, if not downright ridiculous. If time is beating in the gravitational fields of the dwarves, and if these fields are circumambient to these stars, then how does one separate out space, place, and time? I am not denying that their separate measurements can be taken, but the mere measurement of something is not equivalent to its distinctness. In other words, to slide back into the broader question at hand again, when does the link between space, time, and place transform three inseparable entities into one entity with three aspects?

Kaku's writings can be deployed to bolster an argument that space itself has place-like characteristics. Space, according to Kaku, is not the even, smooth, ubiquitously uniform surface we typically imagine it to be. The universe has "a rough texture; there are ripples and lumps in it: This lumpiness can possibly be explained as the ripples from the original big bang, which have been stretched as the universe expanded" (Kaku 2004, 214), making such ripples a sort of combination of the aftereffects of the big bang and the stretch marks of the universe as it expands. It now seems possible that in some parts of the universe, space is twisted, in torsion, antisymmetrical, and, generally, "bizarre" (Kaku 2004, 229). In his *A Different Universe: Reinventing Physics from the Bottom Down*, Robert B. Laughlin concurs with Kaku's assessment: "The ostensibly empty vacuum of space . . . is not empty at all but full of 'stuff'" (2005, 17). Laughlin bases this on the capacity of space to transform matter as space's "sympathetic motion when matter passes by changes the matter's properties slightly, just the way sympathetic motion of the electrons and atoms in a piece of window glass modifies the properties of light as it passes through, causing it to refract" (2005, 17). Our conception of space as a smooth, even surface that has no impact on anything whatsoever may simply be a lingering vestige of a congeries of misconceptions inherited from Newton's notion of a universal and uniform space. And there is nothing particularly new about this way of thinking about space. In a radio lecture delivered in 1948, Maurice Merleau-Ponty comments on this variable aspect of space: "Space is composed of a variety of regions and dimensions, which can no longer be thought of as interchangeable" (2004, 50).

But now to Malibu's Zuma Beach. In their "Multidecadal-Scale Beach Changes in the Zuma Littoral Cell, California," James G. Zoulas and Antony R. Orme state, "Beach erosion [in the Zuma cell] is aggravated when storm waves are superimposed on super-elevated ocean levels and high tides, notably under El Nino conditions" (2007, 282). Without delving into too many technical details, let me simply define the El Nino as a cyclical oceanic-atmospheric condition that has major climatic effects upon the Pacific Ocean and the coastline of North and South America and a cell as a littoral area "with its own sediment sources and sinks" (Zoulas and Orme 2007, 280). The study undertaken by Zoulas and Orme, the results of which were published in *Physical Geography*, is an effort to examine "beach changes in a discrete southern California cell [Zuma] between 1928 and 2002" (2007, 277).

So allow me to parse this examination of Zuma cell beach erosion in terms of space, time, and place. Try an examination of the cell without using the factor of space as a primary component. Of course, one has to omit tides, as the moon and its effects, operational at a distance of some 239,000 miles from the Earth, have been stripped out of this analysis. One may also have to divest the research project of at least a portion of the El Nino effects, as these are the result of an unusual admixture of oceanic and atmospheric conditions, the parsing of which into units either strictly to do with space or place is impossible. Furthermore, since the El Nino is cyclical, time and its relationship to spatial effects cannot be discounted either, as the "prolonged erosion of California beaches has been linked with winter storm seas generated during El Nino phases of the El Nino–Southern Oscillation (ENSO) system that recur at quasi-cyclical intervals of 3–7 years" (Zoulas and Orme 2007, 278). Time, then, abuts this analysis, not only on a daily basis, with the tidal ebb and flow, but also on a seasonal basis with those "winter storm seas," as well as on the basis of the three-to-seven-year ENSO intervals. But I have already noted that time cannot be canceled out, especially when one recalls that Zoulas and Orme are specifically attempting to examine the *multidecadal-scale* Zuma cell beach changes that occurred *between 1928 and 2002*. How can time be stripped out when time is at the very core of this study? But perhaps place, the junior partner of the space-time-place triad, can be subtracted from the project. But then this is the *Zuma cell* these geographers are examining, "a 27-km long segment of coast between Point Magu and Point Dume whose sediment budget is supplied almost entirely from sources within the cell" (Zoulas and Orme 2007, 278). Without this place, or at least *some* place, studies of beach erosion will have a hard time. And so place, space, and time have quickly become enmeshed to a point beyond delinkage.

The more one examines the possibility of thinking through the separation of place, space, and time, the more one realizes that such a possibility is a project riddled with the ludicrous and perforated with the ridiculous, amounting to mere "nonsense on silts," to borrow Bentham's phrase that he used to describe metaphysical blathering. In fact, in my estimation, such a separation is tantamount not only to metaphysical blathering but to blathering of a *physical* kind as well, as is exemplified by my infelicitous attempt to diagnosis beach erosion, the beating of white dwarves, or the red shift without the combination of the three: space, time, and place.

Perhaps I have not made a case formidable enough to transform spacetime into the unit of space-time-place. However, at the very least, I think I can rest assured that I have brought space, time, and place into this structural apparatus, whether as three distinct entities or as the combined entity for which I have been arguing. Space, time, and place must be accounted for in any proper schema of the everyday and, now that at least an attempt has been made to account for them, their corollaries, history and geography, must be brought into the fold.

CHAPTER 3

Time Goes Vertical;
Space Yields In

What I want to do in this chapter is to bring history and geography into this unfolding schema of the everyday. History is brought in through the conveyance of de Certeau's conception of the ruse and geography through Lefebvre's conception of spatial secretion. This will add the diachronic depth of history and the synchronic width of geography into my analysis of the everyday.

The Ruse of the Ruse

"It is de Certeau's theories, and especially his consideration of the potential for resistance within the tactical times of everyday life, that have provided the intellectual space for a more strident populism," writes Roger Silverstone in *Television and Everyday Life* (1994, 162). Setting aside the question of the veracity of this claim, it is this matter that I wish to first consider here: "the potential for resistance within the tactical times of everyday life," especially as represented by the tactic of the ruse. I first outline de Certeau's notion of tactic and strategy, delineate the ruse as a primary instrument of the former, and then illustrate the ruse through the character of Schweyk, "arguably the outstanding fictional figure of our century," as he appears in four manifestations: in the novel by Jaroslav Hasek, *The Good Soldier Svejk*, in *Schweyk in the Second World War*, the play by Brecht, and as that figure is embodied in Hasek's bohemian existence, and by Brecht in his testimony before the House Un-American Activities Committee (Manheim and Willett 1975, xiv). I then ask whether the ruse is an act of rebellion, revolution, or possibly both, before finally situating tactics within the set of everyday practices of everyday life.

De Certeau outlines his schema of strategy and tactics in the introduction to *The Practice of Everyday Life*:

> I call a "strategy" the calculus of force-relationships which becomes possible when a subject of will and power (a proprietor, an enterprise, a city, a scientific institution) can be isolated from an "environment." A strategy assumes a place that can

be circumscribed as *proper* (*propre*) and thus serves as the basis for generating relations with an exterior distinct from it (competitors, adversaries, "clienteles," "targets," or "objects" of research). Political, economic, and scientific rationality has been constructed with this strategic model.

I call a "tactic," on the other hand, a calculus which cannot count on a "proper" (a spatial or an institutional localization), nor thus on a borderline distinguishing the other as a visible totality. The place of a tactic belongs to the other. A tactic . . . has as its disposal no base where it can capitalize on its advantages, prepare its expansions, and secure independence with respect to circumstances. The "proper" is a victory of space over time. On the contrary, because it does not have a place, a tactic depends on time—it is always on the watch for opportunities that must be seized "on the wing." (1984, xix; italics, de Certeau)

But do such tactics have any efficacy? How can they, when de Certeau admits that "whatever" a tactic "wins, it does not keep" (1984, xix)? In other words, none of the gains or victories seized by the use of a tactic can be permanent. Whatever is garnered through the tactics of "clever tricks, knowing how to get away with things, 'hunter's cunning,' maneuvers, polymorphic simulations, [and] joyful discoveries, poetic as well as warlike," evaporates, vanishes, making no dent in the armature of power. For such tactical "practices are exposed upon a field controlled by stronger force. They lack the fortifications [i.e., the space] and the assurances provided by multifarious kinds of capital" (Ahearne 1995, 162). "A tactic is an art of the weak," says de Certeau (1984, 37). Referring to Clausewitz's *On War*, de Certeau claims that "the more a power grows, the less it can allow itself to mobilize part of its means in the service of deception" due to the fact that "power is bound by its very visibility," whereas deception and trickery are "possible for the weak" and are frequently their "only possibility" (1984, 37).

But, clearly, this "seems mistaken," as Ahearne kindly puts it: deception and trickery are not solely the instruments of the weak or the marginalized (1995, 162). One has only to consider what was probably the most adroit operation of the CIA—the overthrow of Guatemalan president Arbenz in 1954—to demolish this thesis. Deft use of radio broadcasts originating from a station "deep in the jungle" (actually originating in the CIA station in Miami); fake obituaries of pro-government politicians repeatedly appearing in newspapers; and general rumormongering of an imminent U.S. invasion were all used to fuse a panic that led to the collapse of a reformist government and the resignation of Arbenz. These operations resulted in a fairly bloodless coup, followed by an extremely bloody aftermath consisting of forty years of military dictatorship. That deception, especially the ruse, may be a favored instrument of the marginalized is quite true. For instance, during the Algerian War of Independence, Kabyle villagers acted "as if they supported the French (in order to receive arms and military training) but in practice aligned with the nationalists" (Eickelman 2009, 263–64). The use of deception is not limited to military and political situations: they occur on an everyday basis, being deployed in numerous situations. Deception and its first cousin, deceit, are used by children trying to avoid punishment and by hedge fund managers attempting to rake in bigger and better profits, and are the very gist of the craft of acting: "Alexandre Dumas *fils* saw her [Sarah Bernhardt's] deceitfulness as an essential part of her genius: 'You know,' he said of the famously thin actress, 'she's such a liar, she may even be fat!'" (Robb 2010, 8).

But can acts of deception, ruses, and other tactics of the weak lead to a reversal and overthrow of the powerful and their strategies? Or do they instead serve as merely acts of a kind of rebellion manqué, a sort of feint to revolution that may satisfy the short-term goal of poking a finger in the eye of the powerful but, to keep with the metaphor, leave the eye intact?

Before attempting to answer those questions, let us consider the opposition of de Certeau's "anti-discipline" with Foucault's "discipline."

> In Foucault's sense power is productive of subjects; it is a microphysics of discourse and practice, multiplied through social space, that evokes positions or configurations of subjectivity. But unlike Foucault, de Certeau will focus on "anti-discipline," on the moments or twisting of subject positions by consumers of technologies. He sees these as resistances that have been undertheorized and therefore marginalized. . . . De Certeau's gesture here is that of . . . rescuing from obscurity a moment of social domination. He reinscribes this moment as one of resistance "from the bottom up." (Poster 1992, 102)

If we accept Poster's analysis, this conforms with Hacking's notions, lining up de Certeau with Goffman's side of the ledger, as working from the anti-disciplinary street-level bottom-up position, while putting Foucault on the disciplinary side of the ledger in the penthouse, top-down position. This may turn out to be much too simplistic and should never be conflated with a notion that Foucault *favored* those who controlled the technologies of power and administered its regimes of discipline. However, de Certeau does seem to offer an egress from, or at least a movement away from, the unyielding dominance of the disciplinary structures that Foucault *seems* to treat as if they are omnipotent, relentlessly colonizing everything in their omnivorous path. According to Ben Highmore, for de Certeau, "Everyday resistance is not seen as the confrontation or contestation of 'discipline,' but simply as that which isn't reducible to it. The everyday is both remainder and excess to such a 'grid.' . . . The activities of everyday life stubbornly evade capture by discipline" (2006, 108, 110). Or as de Certeau himself puts it: "Beneath what one might call the 'monotheistic' privilege that panoptic apparatuses have won for themselves, a polytheism of scattered practices survives, dominated but not erased" (de Certeau 1984, 49). Or as Pierre Bourdieu puts it, in a statement that must have sent Frank and Lillian Gilbreth as well as Frederick Taylor rolling over in their scientifically managed graves: "The most banal tasks always include actions which owe nothing to the pure and simple quest for efficiency" (1990b, 7).

But what are these "other procedures" that, according to de Certeau, are sucking the very blood out of "the system of discipline and control"? Highmore complains that as appropriated by postmodern followers of de Certeau as well as their antagonists, "de Certeau's work seems to have been massively constrained and contained. Exegesis and employment of his work is often caught between a celebratory account of minor acts of 'transgressive' opposition (ripped jeans, fanzines, skateboarding, graffiti, and so on) and the condemnation of such celebration in the name of a more pragmatic politics" (2006, 103). If de Certeau is posing a tactics of antidiscipline meant to match the power of Foucault's regimes of discipline, they better consist of more than the "rebellious" procedures represented by ripped jeans and fanzines, as tactics of anticonsumption are vacuumed up by marketers and turned around as products at a dazzling speed of co-option. Yesterday's anti-hero is today's spokesperson.

However, the array of tactics available or invented by "everyday people" are of quite a higher order than the short list provided by Highmore would suggest, as Highmore himself is quite well aware of: "The impulse behind de Certeau's project is unequivocal," states Highmore before citing de Certeau: "'If it is true that the grid of "discipline" is everywhere becoming clearer and more extensive, it is all the more urgent to discover how an entire society resists being reduced to it'" (2006, 108). While many may claim that an entire society has already been reduced to a compliant congruence with such a grid, de Certeau "offers, not a counter-narrative, but a para-narrative (and a para-archive)" to such a view, "where everyday life . . . articulates moments of cunning and stubborn resistance" (Highmore 2006, 109).

In order to attempt to clarify what such a para-narrative may look like as well as to gauge whether such a narrative may be a worthy counterpoint to regimes of discipline and their discourses, consider the genealogy of Schweyk. And perhaps the best, as well as the most proper, place to begin that genealogy is with a brief narrative of the life of Czech novelist Jaroslav Hasek, the founder and proprietor of the "Cynoloigcal Institute" (a dog-fancier's shop), a Soviet *apparatchik*, an anarchist as well as the founder of the Party of Moderate and Peaceful Progress Within the Limits of the Law, a vagabond, the editor of *Animal World*, and a common drunk. Born in Prague in 1883, Hasek displayed a predilection for the ruse from an early age: "Not only was Hasek a true bohemian: he was a born mischief-maker and hoaxer as well" (Parrott 1974, viii).

Before turning to Hasek's novel, I cannot resist including the biographical detail that Hasek was fired as the editor of the journal *Animal World* for composing and publishing stories about completely fantasized animals. And I should also add that Hasek fought on the Galician front in World War I, was captured by the Russians and incarcerated in a prisoner-of-war camp, joined the Bolsheviks after the Revolution, rose to the rank of "Secretary of the Committee of Foreign Communists in the Russian town of Ufa," then became "Secretary of the Party Cell of the printing office of the Red Arrow, and a year later [1920] Head of the International Section of the Soviet Communist Party School," prior to returning to Prague to participate anew in "the orgies and excesses of his youth," dying on "3 January 1923" at the age of thirty-nine (Parrott 1974, xiii–xiv).

Hasek's Svejk is a good-natured self-confessed "imbecile" who is perpetually seeing the sunniest side of even the direst of circumstances: "I was discharged from the army for idiocy and officially certified by a special commission as an idiot. I'm an official idiot" (Hasek 1974, 20). Having once landed in "the lunatic asylum" after being diagnosed as being "a patent . . . idiot according to all the natural laws invented by the luminaries of psychiatry," Svejk recalls his time there "in exceptionally eulogistic terms":

> I really don't know why those loonies get so angry when they're kept there. You can crawl naked on the floor, howl like a jackal, rage and bite. If anyone did this anywhere on the promenade people would be astonished, but there it's the most common or garden [variety] thing to do. There's a freedom there that not even Socialists have ever dreamed of. . . . Everyone there could say exactly what he pleased and what was on the tip of his tongue, just as if he was in parliament. Sometimes they used to tell each other fairy stories and started fighting when something very bad happened to the princess. The wildest of them all was a gentleman who pre-

tended to be the sixteenth volume of Otto's Encyclopedia and asked everyone to open him up and to find the entry: "Cardboard box stapling machine," otherwise he would be done for. ..., There was one well brought-up inventor ... who continually picked his nose and said only once a day: "I've just discovered electricity." As I say it was very pleasant there and those few days which I spent in the lunatic asylum are among the loveliest hours of my life. (Hasek 1974, 31–32)

Well now, perhaps this doesn't comport so neatly with our highly developed sense of political correctness, but Svejk's ability to toss a silver lining around the most foul of circumstances reflects something of a practice that, as Highmore points out, has too often been ignored in the academic analyses of de Certeau's work: the capacity to stubbornly endure through all and every variety of hardship. Highmore locates such practices primarily among the right. However, such obdurateness doesn't necessarily exclusively adhere to those who are conservatives: such a tendency could just as well be manifested in those on the left who may cling to memories of a more communal era prior to the rigors of globalization, whether such memories reflect a "true" state of affairs or not. As Michael Herzfeld puts it when enunciating a term he coined, "structural nostalgia," the phrase is intended to denote "a collective representation of an edenic order—a time before time—in which the balanced perfection of social relations has not yet suffered the decay that affects everything human. Structural nostalgia characterizes the discourse of both the official state and its most lawless citizens," as well as, I think can safely be presumed, potentially everyone in between, whether of the left or the right, at least at certain times and in certain places (2005, 147).

But the main point I want to make here is that Svejk's stubborn sunniness is a tactic of survival; it is a tool that he uses to combat any and every tight spot in which he finds himself: having landed in a jail cell, for instance, he affably states with all due appreciation and affability: "It's not too bad here. ... They've at least planed the wood on this plank-bed" (Hasek 1974, 37). This obdurate optimism can be conceived of as an everyday practice, a survival technique, a tactic deployed to withstand the rigors of stretches in jail, nine-to-five jobs, military service, domestic chores, university seminars, family life, and so on. Svejk's transformation of the degraded into the durable and even delightful is part of the necessary "wit" employed "in our continual reinvention of everyday life," a wit required for perdurance through the disparate ordeals of life (Conley 1992, 47).

The officials and authorities are not always convinced of Svejk's imbecility: some of them even suspect that it might be a wily ruse: when hauled before a commission of military doctors attempting to diagnose his suitability for service, they "were remarkably divided in their conclusions about Svejk. Half of them insisted that Svejk was a 'half-wit,' while the other half insisted he was a scoundrel who was trying to make fun of the war" (Hasek 1974, 76). This indecision may be perfectly understandable, given that Svejk seems to be tightrope-walking that fine line between idiocy and brilliance, a line played to perfection by many an "imbecile," such as Lear's Fool, Harpo Marx, and Jacques Tati's Monsieur Hulot. The commission's confusion about Svejk's mental status may also be understandable if they had had access to accounts of Svejk's "treatment" prior to his appearance before the commission, a treatment including repeated stomach pumping and twice-daily enemas, which Svejk manages to endure with steadfastness, even offering up words of encouragement to the medical corpsman who is administering said treatments: "Don't spare me," he invited the myrmidon who was giving him the enema. "Remember your oath. Even if it was your father or your own brother who was lying here, give him an enema without batting an eyelid. Try hard to think that Austria rests on these enemas and victory is ours" (Hasek 1974, 69). That final line vaults past unreflective idiocy into intentional satire, no matter how innocent and imbecilic Svejk may have seemed before the delivery of these "encouraging words."

Much of the nature of the tactical ruses employed by Svejk and his fellow "malingerers" has to do with the tricks necessary to avoid military service, tactical devices familiar to any American male who came of draft age during the Vietnam War. The shamming of insanity or epilepsy, muteness or blindness, consumption or pneumonia, diabetes or typhus; and the auto-dislocation of one's own limbs, the self-administration of various poisons, and the employment of "experts" to induce fevers and assorted other disabling maladies are all tactics considered or practiced by Svejk and his colleagues.

> "I know a chimney-sweep in Brevnov," remarked another patient. "For ten crowns he'll give you such a fever that you'll jump out of the window." "That's nothing," said another. "In Vrsovice there's a midwife who for twenty crowns will dislocate your leg so well that you'll be a cripple until your death." ... "The best thing to do," explained somebody from the door, "is to inject paraffin under the skin of your arm. My cousin was so fortunate as to have his arm cut off under the elbow and today he has no trouble for the rest of the war." (Hasek 1974, 64–65)

To claim that these are revolutionary tactics would be to make a claim that goes beyond the evidence, but to make a claim that such tactics serve as fillips for the undermining of regimes of discipline while also, and primarily, serving to increase the chances of one's own survival seems to fall well within the mark.

However, the narrator of *The Good Soldier Svejk* does occasionally make statements that could be interpreted as having a much more revolutionary overtone. For instance, chapter 9 of the novel, entitled "Svejk in the Garrison Gaol," begins with the following: "For people who did not want to go to the front the last refuge was the garrison gaol. I once knew a probationary teacher who was a mathematician and did not want to serve in the artillery and shoot people. So he stole a lieutenant's watch to get himself into the garrison gaol. He did this deliberately. War neither impressed nor enchanted him. Shooting at the enemy and killing with shrapnel and shells equally unhappy probationary teachers of mathematics serving on the other side seemed to him sheer idiocy" (Hasek 1974, 79). Too much thinking along these lines and warmongers would have no fodder, an outcome with revolutionary implications, fraught with peril for those determined to "protect the homeland," "civilize the natives," "spread democracy," or "advance the universal blessings of civilization."

Malingering as a tactical device to sabotage the order of the everyday carries within it an alternative version of that order. The malingerer as con artist must give the impression of at least appearing to submit to the set of dominant norms while also simultaneously sustaining an alternative set of norms within the subordinate order, a double-hinged ruse calling for extreme acuity. This is something acknowledged by Goffman, a student of con artists, as he recognized that besides mastering their own set of norms, "successful con artists understood the interaction order as well as any sociologist" (Fernandez 2003, 209).

Moving from Hasek's novel to Brecht's play, *The Good Soldier Svejk* to *Schweyk in the Second World War*, first it should be noted that Brecht's adaptation was written in the United States, while Brecht was on a global run from the Nazis, traveling from Berlin to Copenhagen to Stockholm to Helsinki to Moscow to Vladivostok before sailing to the United States, where he split his time between New York City and Los Angeles, with the majority of time spent in the latter amid the famous German exile community of Los Angeles, which included the film director Fritz Lang and the actor Peter Lorre, the philosophers Theodor Adorno and Max Horkheimer (both of whom Brecht loathed for their "salon Marxism" while they, in turn, loathed him for his "vulgar" and "undialectical" Marxism), and the novelists Lion

Feuchtwanger, Alfred Doblin, and Thomas Mann (Lyon 1980, 295, 257–258). Brecht had long been interested in dramatizing *The Good Soldier Svejk* "so [that] people can see the ruling forces up top with the private soldier down below surviving all their vast plans" (Mannheim and Willett 1975, xv). In fact, he had served as a dramaturge on an adaptation of the play by Max Brod and Hans Reimann that had been produced under the direction of Erwin Piscator in Berlin in 1928; Brecht set to work on his own version in 1943: on April 3 of that year "a mixed program at Hunter Collège" was produced by "Ernst-Josef Aufrichr, the former Berlin impresario who had first staged *The Threepenny Opera* in 1928 and was now in New York after escaping from Unoccupied France" (Mannheim and Willett 1975, xv). During this "mixed program," Kurt Weill and Lotte Lenya performed some of Brecht's songs, "including 'Und was bekam das Soldaten Weib?' which Weill had recently set; the program finished with a turn by the Czech clowns George Voskovec and Jan Werich entitled 'Schweyk's spirit lives on,'" which seems to have been a kind of herald or harbinger of the final form of Brecht's play (Mannheim and Willett 1975, xv).

The title, "Schweyk's spirit lives on," orients us once again to an emphasis on survivability as the prime element of those such as Schweyk who use the ruse and indeed are masters of this tactic. The ruse is a tool to ensure that "one lives another day," as it were: clearly this is the case with Hasek's depiction of the Czech malingerer and is also the case with Brecht's Schweyk. However, before jumping into any further analysis of *Schweyk in the Second World War*, I'd like to briefly cross swords with Highmore's analysis and disagree with his sharp distinction between the ruse and the kind of stubborn perdurance that can also be viewed as a tactical device of resistance. The ruse, it seems to me, can be conceived as part of such stubbornness while obdurate endurance can manifest itself in a ruse. The cunning of a fox and a kind of doughty muleheadedness are not necessarily mutually exclusive: in fact, either one can be thought of as part and parcel of the other, a kind of fantastical two-headed creature "keeping us keeping on," as it were. Such a dual operation can be perceived in the examples of "malingering" noted above: cunning is certainly required to pull off such a tactic, but also an unmitigated and redoubtable sense of stubbornness as well, especially if the ruse needs to be maintained for weeks and months or even years at a stretch.

Brecht's Schweyk also reflects this kind of dual nature, exhibiting both the cunning of the fox and the endurance of the mule while also being quite willing to be as subservient as a whipped dog, if that is what is needed to survive another day. In contradistinction to the clever ploy of answering that he "pisses yellowish-green," when he is asked by the SS if he pisses yellow or green, which is simultaneously an obsequious response and an artful ruse, when asked by an SS man if he shits thin or if he shits thick, Schweyk replies, "Beg to report, Herr Platoon Leader, I shit the way you want me to" (Brecht 1975, 77). He uses the same tactic of complete and utter compliance when an SS man is attempting to force him to confess to making seditious remarks: "If you want me to confess, Your Excellency, I'll confess, it can't hurt me. But if you say, 'Schweyk, don't confess,' I'll talk myself out of it till they tear me to pieces" (Brecht 1975, 78). There are two points to be made in regard to these remarks: one is that nothing can destroy Schweyk: "it can't hurt me." His "imbecilic" near-preternatural vitality and his instinct for survival can withstand anything, even a deployment to the Russian front where he meets a disoriented Hitler, lost in the snow outside of Stalingrad. It is this capacity to survive through any and every hardship that at least partially explains Schweyk's odd heroic quality while also giving him his status as the quintessential Czech, the dog "hustler," the greatest of underdogs. Secondly, this citation points out the essential irrationality of torture or its latest manifestation, "enhanced interrogation techniques": Schweyk is simultaneously willing to confess to anything and *not* willing to confess to anything, whatever "Your Excellency" desires, a double-edged sword that obviously is very likely to lead to information of no value whatsoever, that is, *mis*information.

When discussing a recent assassination attempt on Hitler, Schweyk manages to get one in edgewise by stating that Hitler could not be replaced "with any old idiot," implying, of course, that an *uber* idiot would be needed to replace the idiocy par excellence of Der Fuhrer (Brecht 1975, 75). When accused of being a black-marketer, Schweyk remarks, "There's got to be order. Black-marketing is an evil and it won't stop until there's nothing left. Then we'll have order right away, am I right?" which, on the face of it, is a statement of compliance while its thinly disguised subtext can be read as being a simple statement of fact as well as a dire prediction (Brecht 1975, 109). Schweyk as everyman, as working-class hero, also serves as a moral prop for those around him who do not share his unmitigated "idiotic" optimism: "Don't ask too much of yourselves," he advises his fellow Czechs. "It's quite a job just being around nowadays. Keeps you so busy just staying alive, you haven't much time for anything else" (Brecht 1975, 90).

This focus on survival and the compassion Schweyk manifests for others as they do their utmost to live to see another day must have struck a mighty chord with Brecht: the very fact of the war itself, the journey with his family nearly halfway round the world to the alien environment of the United States in one of its most alienating environments, Hollywood; the death of his mistress and collaborator, Margarette Steffin, who died of tuberculosis in Moscow in 1941 while also in flight from Nazi Germany; the suicide of his close friend, Walter Benjamin, in 1940; and his own difficulties making a living while suffering the indignities of hustling scripts in Hollywood—all must have contributed to his sense that in the character of Schweyk was embodied the very nub of the sense of survival that was needed in these, the darkest and most desperate of times.

Though it seems that Brecht himself did not typically manifest the ploys and ruses that we might call Schweykian, as the playwright was more noted for his brutal candor than for the ambiguous slyness of the Prague dog hustler, there did come a time when Brecht seems to have deployed a sort of Schweykian discourse and for much the same reason, survival, to live to see another day, and hopefully, not behind bars.

In October 1947, Brecht, along with a small legion of Hollywood screenwriters, producers, directors, and one "lone actor," who came to be known collectively as the "Hollywood Nineteen," received a summons to appear before the House Un-American Activities Committee (HUAC) (Lyon 1980, 318). Desperate at the time to return to Europe, with visas in hand for himself and his family and transportation already arranged, Brecht was not about to delay his departure by making foolhardy statements before the committee. Though the rest of the Hollywood Nineteen planned to confront the committee head-on, no holds barred, Brecht insisted on a different strategy, as he thought "that their *tactics* were wrong" (Lyon 1980, 319; italics, mine). "Certainly Brecht recognized the Committee as a threat to individual freedom. He himself planned to oppose it through an outward show of cooperation that was really a brand of cunning. Outsmarting a powerful enemy was to him a valid form of opposition. He considered martyrdom to be folly in any political struggle, and the eighteen [the rest of the Hollywood Nineteen] had adopted a course that smacked to him of precisely that" (Lyon 1980, 39).

In other words, Brecht was planning to take a Schweykian stance in front of the committee: "in Schweyk" he had signaled an approval for "an unheroic brand of behavior designed to do nothing more than save one's own skin," an end that Brecht accomplished without risking anyone else's reputation. This matches Brecht's tactics when dealing with the Communist apparachitniks in East Germany in the 1950s: "cheerful compliance, even subservience, in his dealings with the authorities on the one hand, stubborn persistence in his own essential convictions on the other" (Esslin 1980, 164).

Brecht rehearsed for his appearance in front of HUAC with his attorneys, Bartley Crum and Robert W. Kenny, the political journalist Hermann Budzislawski, and film director Joseph Losey. While prepping with Budzisawski, "they devised an answer to whether Brecht had written a given poem. He would reply that he had not written this poem, among other reasons because it was in English, and was quite different from the German poem on which it was based. He . . . used this *ploy* effectively" (Lyon 1980, 323; italics, mine). While rehearsing with Losey, the director "advised Brecht to smoke a cigar during the hearings. J. Parnell Thomas [Chairman of HUAC and a Republican congressman from New Jersey] was an avid cigar smoker, and Losey felt that Brecht might be treated more sympathetically by a fellow cigar smoker. The dramatist added this to a routine he was still rehearsing" (Lyon 1980, 328). None of this will be surprising to anyone with even the slightest knowledge of the workings of the justice system, as attorneys rehearse with their clients and frequently employ ploys and ruses for the benefit of a "fair" outcome. However, this is precisely the point: ploys, ruses, and the development of a routine are all everyday methods, if not of determining the outcome of events, at least of influencing them. That they can be Schweykian in character does not, of course, mean that only those with a knowledge of Hasek's *Svejk* or Brecht's *Schweyk* have access to their performance: the performance of this type of routine is, well, routine.

Brecht's intensive preparation served him well. He was even given the ultimate compliment by the HUAC chairman, J. Parnell Thomas, at the conclusion of his testimony: "Thank you very much, Mr. Brecht. You are a good example to the witnesses of Mr. Kenny and Mr. Crum" (Bentley 1971, 220), which, of course, is more than a tad Schweykian as well, as Brecht may have been the most revolutionary of all the show business witnesses to appear before the committee and yet was being held up as an example to the remaining eighteen of the Hollywood Nineteen. One of Brecht's ploys was to screen himself off from the more confrontational tactics of the other witnesses by claiming that as a mere guest in the United States he was compelled to be compliant in his testimony. Instead of defying the committee in the manner of the actor Lionel Stander or the writer John Howard Lawson, Brecht would be fully cooperative *and* totally elusive, a difficult trick to pull off, requiring a certain facile agility that Brecht clearly displays in his testimony. "Brecht himself was only too eager to put his impertinence and Schweykian servility against what he considered the darkest and most evil forces in the country. He had always enjoyed such encounters and delighted in misleading pompous representatives of authority by 'sticking strictly to the untruth'" (Esslin 1980, 70). Only in this way could he avoid both the martyrdom of a Lawson, who was blacklisted and spent a year in prison for contempt, or the surrender of an Elia Kazan, who dealt out names to the committee like a drunken faro dealer and quickly thereafter signed a $500,000 contract with one of the Hollywood studios.

In the following exchange between Brecht and Robert Stripling, chief investigator for the committee, Brecht manages to conflate his work, which is clearly meant as an indictment of capitalism and its corollary, liberal democracy, as merely intended to aid in the overthrow of Nazism.

MR. STRIPLING: Mr. Brecht, is it true that you have written a number of very revolutionary poems, plays, and other writings?

MR. BRECHT: I have written a number of poems and songs and plays in the fight against Hitler and, of course, they can be considered, therefore, as revolutionary because I, of course, was for the overthrow of that government. (Bentley 1971, 209–10)

And when Stripling pressed further and asked, "How many of your writings been [sic] based upon the philosophy of Lenin and Marx?" Brecht, who "since 1930 . . . had openly based all his writings on a Marxist viewpoint . . . replied with complete confidence in the ignorance of the Committee: 'No. I don't think that is quite correct. But of course I studied; I had to study as a playwright who wrote historical plays; I, of course, had to study Marx's ideas about history. I do not think intelligent plays today can be written without such study. Also history now, written now, is vitally influenced by the studies of Marx about history'" (Esslin 1980, 72).

When questioned about *The Measures Taken*, one of Brecht's early plays in which a revolutionary worker is executed after mangling a political assignment, Brecht fends off Stripling's line of attack by claiming that the play is merely an "adaptation of an old religious Japanese play, called [a] Noh play, and follows quite closely this old story which shows the devotion for an ideal until death" (Bentley 1971, 211). When pressed by Stripling if the play is "pro-Communist" and "whether or not one of the characters in this play was murdered by his comrade because it was in the best interest of the Communist Party," Brecht deflects this with an approach that can be classified as both insistent and compliant.

MR. BRECHT: No, it is not quite according to the story.

MR. STRIPLING: Because he would not bow to discipline he was murdered by his comrades, isn't that true?

MR. BRECHT: No, it is not really in it. You will find, when you read it carefully, like in the old Japanese play where other ideas are at stake, this young man who died was convinced that he had done damage to the mission he believed in and he agreed to that and he was ready to die, in order not to make greater such damage. So he asks his comrades to help him and all of them together help him to die. He jumps into an abyss and they lead him tenderly to that abyss. And that is the story.

MR. CHAIRMAN: I gather from your remarks, from your answer, that he was just killed, he was not murdered.

MR. BRECHT: He wanted to die.

MR. CHAIRMAN: So they kill him?

MR. BRECHT: No, they did not kill him—not in this story. He killed himself. They supported him, but of course they had told him that it were better when he disappeared, for him and them and the cause he also believed in. (Bentley 1971, 212)

This parsing of the "measures taken" is a careful transformation of the rigorously lethal discipline of Soviet-style Communism into a tender neo-communal leap into the abyss. Such a parsing does not quite align with Hannah Arendt's gloss of the play: "For the measure taken is the killing of a Party member by his comrades, and the play leaves no doubt that he was the best of them, humanly speaking. Precisely because of his goodness, it turns out, he had become an obstacle to the revolution" (1983, 241). When placed against his apologias for Stalin's reign of terror, Brecht's hermeneutics can be read as a feat of legerdemain, the misdirection of a wily conjurer diverting his audience from the "trick" behind the trick, which is hidden in plain sight; James Lyon in *Bertolt Brecht in America* cites conversations with Hans Viertel and Egon Breiner, close friends of Brecht's, to show that Brecht's support of Stalin extended to a justification of the show trials:

By the time he [Brecht] reached America, he had developed a battery of sophisticated arguments to show that these trials were necessary, and they made him sound like an apologist for Stalinist terror. Viertel and Breiner . . . recall that their most violent disagreements centered on this topic. Viertel remembers Brecht's statement that Stalin had no way of making ninety million illiterate muzhiks understand why Bukharin and others were ideologically wrong; he had to treat these men as criminals to make his point clear. (Lyon 1980, 295–96)

That Brecht's sympathy, and even empathy, with *the measures taken* by Stalin may have extended to executions on a massive scale can be viewed as possible (and even probable).

But there's more to this exchange than meets the eye, as Brecht, having become convinced that the committee's "investigators" knew next to nothing about his work, was not only deflecting and diverting Stripling's attack: he was also lying through his teeth, as the playwright "intentionally confused" *The Measures Taken* "with the earlier play *Der Jasager*, which *was* based on a Japanese legend and had been a kind of preliminary study for" *The Measures Taken* (Esslin 1980, 71; italics, Esslin). Of course, if Brecht had been caught in the lie, he had a way out of the tangle he had deliberately created: "If challenged, Brecht could always have excused himself with having simply confused two of his plays that have much in common. But the Committee, of course, did not know the facts" (Esslin 1980, 71). The deft lie, along with its essential backups, the robust explication, the deftly delivered excuse, and the circuitous but believable rationale, should be added to our catalog of ruses. But, of course, total and complete ignorance on the part of one's opponents also helps in such situations.

Another ploy that Brecht used while testifying was to deflect criticism by throwing the blame on the translation, or rather, the *mis*translation of his work.

MR. STRIPLING: Did you collaborate with Hans Eisler on the song "In Praise of Learning"?

MR. BRECHT: Yes, I collaborated. I wrote that song, and he only the music.

MR. STRIPLING: Would you recite to the Committee the words of that song?

MR. BRECHT: Yes, I would. May I point out, that song comes in another adaptation I made, of Gorky's play, *Mother*. In this song a Russian worker woman addresses all the poor people. (Bentley 1971, 217)

Having recruited the same ploy used while defending *The Measures Taken*, that the work in question should be viewed as a mere adaptation instead of an original composition, Stripling recites the lyrics of "In Praise of Learning" and then Brecht sets his next ruse into action:

MR. STRIPLING: Learning now the simple truth.
You, for whom the time has come at last.
It is not too late.
Learn now the A, B, C,
It is not enough, but learn it still.
Fear not, be not downhearted.
Begin, you must learn the lesson
You must be ready to take over.

MR. BRECHT: No, excuse me, that is the wrong translation. That is not right. (*Laughter.*) Just one second, and I will give you the correct text. (Bentley 1971, 94, 217)

So here Brecht uses humor, one of the most effective tactics of the under-dog, to undermine the line of questioning being pursued by Stripling. The source of humor is a bit obscure and may have had something to do with Brecht's manner or perhaps his tone of voice. Whatever its source, the ploy apparently hit the mark, as laughter ensues. The "correct" text is then offered up by Brecht's interpreter, a Mr. Baumgardt, on loan to the committee from the Library of Congress: according to Baumgardt, the last line should read "taking the lead" instead of "to take over." The interpreter's intervention is followed by a comment from Chairman Thomas: "I cannot understand the interpreter any more than I can the witness" (Bentley 1971, 218). Obfuscation through garbled speech or the use of a thick accent, whether that accent be "real" or exaggerated for effect, are ruses manipulated here to Brecht's ben-efit. Commended for the candor of his testimony, even though members of the committee (or at least its chairman) evidently had some difficulty even comprehending what was being uttered by the witness, the Committee re-leased the playwright without reproach. Brecht left the next week for Europe, never to return to the United States.

Let me make a lateral move here to the infamous Army-McCarthy hear-ings and their undoing by the attorney for the U.S. Army, Joseph Welch, in order to illustrate what I understand as the difference between the tactical and the strategic in de Certeau's schema. "Strategies," writes de Certeau in *The Practice of Everyday Life*, "are actions which, thanks to the establishment of a place of power ... pin their hopes on the resistance that the *establishment of a place* offers to the erosion of time," while tactics "depend on a clever *utili-zation of time*" (1984, 38–39; italics, de Certeau). Tactics, then, rely on a sense of timeliness, not a control of place or space. Brecht, for instance, occupies a tactical position in front of HUAC and effectively uses his comic sense of timing to divert the committee's examination, whereas Stripling occupies a place of power with all its strategic legal resources: the power of subpoena, the right to indict, and so on. Strategies "require an accumulated financial, symbolic and/or scientific 'capital,' together with a corresponding measure of security and stability," while tactical "practices are exposed upon a field controlled by a stronger force," a description certainly aligning with the tactics practiced by Svejk, Schweyk, and Brecht; tactics "lack the fortifications and the assurances provided by multifarious kinds of capital" (Ahearne 1995, 162).

When Welch lambastes Senator Joseph McCarthy with the famous words, "You've done enough. Have you no sense of decency, sir? At long last, have you left no sense of decency?" the words do their work not only because they are delivered with passion, conviction, intelligence, and daring, but also because they are invoked by an attorney for the U.S. Army, certainly an entity that has the "accumulated financial, symbolic and/or scientific 'capital,' together with the corresponding measure of security and stability" required for a strategy to succeed. Imagine Brecht suddenly proclaiming sentiments of a similar nature in front of HUAC, and what one has imagined is precisely the strategy used by the rest of the Hollywood Nineteen. For example, the screenwriter John Howard Lawson roared at Thomas: "I am not on trial here, Mister Chairman. This Committee is on trial here before the American people" (Bentley 1971, 154). This is all well and good, and I, for one, hardily agree with Lawson's sentiments; however, he is making a strategic maneuver instead of a tactical one: he is not in the position (or the place) to put the committee on trial; neither does he have the accumulated financial or scientific capital nor the corresponding measure of security and stability to make such a charge stick even if he were in the position to make the charge. Notice that the symbolic has been omitted from this list, as this seems to be what Lawson is banking on with his intrepid frontal assault: that the symbolic weight of what he is invoking will somehow overwhelm the legal (or extralegal) capital subtending the committee. On the other hand, Brecht knew he was in a tactical, not a strategic, position, and so he deployed the tactical tools of the powerless: the ruse, the prank, the ploy, the fabrication, the diversion, the joke, the deliberate act of misdirection, the "Rabelaisian fart in a ceremony commemorating history" (de Certeau 1988, 314). In its most contemporary form, there is also the psych or the punk, as in "the psych-out" or "being punked" by someone.

The acts of pilfering described by James C. Scott in "Everyday Forms of Peasant Resistance" should also be included here. As Scott elucidates the situation of the Malaysian peasant, "such forms of resistance [such as pilfering and evading taxes] are the nearly permanent, continuous, daily strategies of subordinate rural classes under difficult conditions" (1986, 22). Akin to Schweyk, when these peasants perform acts of everyday resistance, they are just as concerned with long- and short-term survival as they are in performing acts of outright rebellion: "The South-east Asian peasant who hid his rice and possessions from the tax collector may have been protesting high taxes, but he was just as surely seeing to it that his family would have enough rice until the next harvest" (Scott 1986, 23). Christine Pelzer White adds that many of the acts of everyday resistance among Vietnamese peasants, such as "instances of literal and figurative footdragging . . . seem to have actually functioned primarily as delaying tactics in an inexorable process of peasant loss of land" (1986, 54).

Let me make a few more remarks about tactics and the everyday before turning to Lefebvre. First, though I have used what might be termed a mini-genealogy of Schweykian qualities to explicate this term, I do not want to give the impression that such tactics are unusual or confined to idiosyncratic situations. They are deployed every day in a plethora of everyday situations, from high school students shamming sickness to workers fudging time cards to wives and husbands covering their tracks during love affairs to power brokers engaging in hyperbole of all kinds to advertisers transforming superfluous wants into requisite needs. Though not essentially instruments of revolution or even of rebellion, they can certainly be employed for such ends. They also can be used to puncture and perforate norms girding everyday situations, allowing some breathing room into disciplinary regimes that otherwise would be suffocating. But, as witnessed through the examination of Schweykian practices, they can also be used to evade and elude authority while doubling as a kind of reproductive instrument to ensure survival.

Through this exploration of the ruse, a sense of history has been inserted into this study, as de Certeau's tactic was examined through the deployment of the ruse in the cases of the lives of Hasek and Brecht and, in turn, their deployment of the ruse in its insertion into their character of Svejk and Schweyk. And what does the historical do for the everyday? The historical gives it depth. Without the historical, the everyday is a veneer, a surface, and therefore superficial. With it, everyday actions are geared into that which rises up from the past and assumes the form of the present. And now it is *time* for space.

Add in Space

Just as the verticality of time (the depth of time) was added, the horizontality of space (the width of geography) is now added. This is done by inserting Henri Lefebvre's notion of sociospatial secretion into my astructural structure. A spatial analysis can aid in an explication of that which surrounds us: the circumambient is made transparent as spatial secretion is elucidated.

"The spatial practice of a society *secretes* that society's space," says Lefebvre in *The Production of Space*; "it propounds and presupposes it, in a dialectical interaction; it produces it slowly and surely as it masters and appropriates it" (1991, 38). Lefebvre foresaw the difficulties of analyzing spatial secretion due to disciplinary fragmentation and the resulting fragmentation of the theorizing of space and its practices. "The dominant tendency [of studying space and spatial practices] fragments space and cuts it up into pieces" (Lefebvre 1991, 89).

To counter such a tendency, spatial secretion must be studied holistically, as geographical features cannot be studied separately from their economy, history, society, politics, and so on. According to Lefebvre, works of fiction that have been composed *about* or *in* the secreted space should also be included in spatial analysis. So, for example, a spatial analysis of New York City should include a spatially inflected analysis of the work of such writers as James Baldwin, Kathy Acker, Edith Wharton, Joel Rose, Norman Mailer, Chester Himes, Henry James, Philip Roth, Grace Paley, Saul Bellow, Henry Miller, and Isaac Bashevis Singer, among others.

It must also be noted that embedded within Lefebvre's conception of space is a Marxian construal of space as a means of production and as forms of use and surplus value or, rather, a Marxian-Hegelian construal, for in Lefebvre's conception of Marxism there is more than a touch of Hegel's analytic tossed in for good measure. "Space is a means of production: the network of exchanges and the flow of raw materials that make up space also are determined by space" (Lefebvre 1979, 287). Space, then, is both that which does the determining and that which is determined. Such is the import of space to Lefebvre that it can apparently perform this double duty. How could that be possible, this doubling up of space as that which does the determining and that which is determined?

In order to comprehend such a possibility of space being both the active agent and the passive patient of this productive process, we must first understand that by "space" Lefebvre means *social* space and that "(Social) space is not a thing among other things, nor a product among other products" (1991, 73). Well, what is it then? In *The Production of Space*, Lefebvre answers this by initially casting social space in a Hegelian dialectal mold: "It [space] subsumes things produced, and encompasses their interrelationships in their coexistence and simultaneity—their (relative) order and/or (relative) disorder" (1991, 73). So subsumption, that simultaneous retention *and* elimination that Hegel terms *Aufhebung* and that is usually translated into English as either "sublation" or "overcoming," is that "mysterious something" that creates itself while destroying itself, maintains itself while shattering itself, preserves itself while obliterating itself.

In *The Production of Space*, Lefebvre asks if the canals and streets of Venice are a work or a product, and then answers his own question with the Hegelian response that they are both an ongoing work *and* a product of that work. "Here, everyday life and its functions are coextensive with, and utterly transformed by a theatricality as sophisticated as it is unsought" (Lefebvre 1991, 74). In other words, the canals and streets of Venice are functioning as everyday transportation routes. Yet they also are *products*, in this case, products with a great degree of theatrical, aesthetic, economic, and productive value, which as products translate into a great deal of surplus value for Venice and its residents. So that the streets and canals of Venice are able to maintain this sociospatial equipoise of work and product through functioning simultaneously as both work (transportation) and product (tourist attraction).

Lefebvre, being Lefebvre, is not content to let it rest at that: "All the same, every bit of Venice is part of a great hymn to diversity in pleasure and inventiveness in celebration, revelry and sumptuous ritual. . . . There is even a touch of madness added for good measure" (1991, 77, 74). Lefebvre is a thinker who, much like the everyday itself, cannot be held captive, eluding our grasp as he continuously persists in staking out new and (sometimes) contradictory material in his analyses. So that, with any given analysis of any given conception, the stipulation must be submitted that any given analysis of any given conception will end up falling short of any conclusive resolution, as the goalpost marking such a resolution shifts through Lefebvre's own feints and parries down the field.

What Lefebvre is suggesting is that a way of studying spatial production (or the production of things in space) needs to be constructed "which would analyze not things in space but space itself, with a view to uncovering the social relationships embedded in it" (1991, 89). Or, as M. Gottdiener puts it in *The Social Production of Urban Space*: "Urban science in general rests on a basic premise that the spatial patterns of settlement space correspond to the action of deep-level forces of social organization," a judgment that can be accepted if we stipulate that "settlement space" corresponds to all built urban space, including office buildings, government institutions, museums, parks, roads, homes, warehouses, stores, and so on (1985, 8). "If a qualitatively new form of space has developed . . . this implies that the very mode of social organization has changed" (Gottdiener 1985, 8). Such a method would "help us to grasp how societies generate their (social) space[s]" (Lefebvre 1991, 91); for instance, Lefebvre asks us to "consider a house":

> The house has six storeys and an air of stability about it. One might almost see it as the epitome of immovability, with its concrete and its stark, cold and rigid outlines. . . . Now, a critical [sociospatial] analysis would doubtless destroy the appearance of solidity of this house, stripping it . . . and uncovering a very different picture. In the light of this imaginary analysis, our house would emerge as permeated from every direction by streams of energy which run in and out of every imaginable route: water, gas, electricity, telephone lines, radio and telephone signals, and so on. Its image of immovability would then be replaced by an image of a complex of mobilities, a nexus of in and out conduits. (1991, 92–93)

We can, Lefebvre tells us, extrapolate this analytic structure out to the street in front of the house, and then beyond to the city, and then beyond to the governance of the city and of the state itself, and so on. However, Lefebvre warns us that there is an "error—or illusion—generated here," consisting "in the fact that when social space is placed beyond our range of vision in this way, its practical character vanishes and it is transformed" into "fetishized abstract space" instead of "space as directly experienced," a space (or spaces) in which people are living, working, sleeping, thinking, dying, making love, and so on (1991, 93). Wrenching space away from its abstract monetized commodity form and inserting it into its use value/everyday form is Lefebvre's primary goal, as he perceives this as forming the plexus of revolutionary praxis. "For Lefebvre, the revolutionary transformation of society requires the appropriation of space, the freedom to use space, the existential right to space (*le droit à la ville*) to all be reasserted through some radical version of sociospatalial praxis" (Gottdiener 1985, 128). Here, as initial exemplars of this, the events of May 1968 could be cited as well as the passage of the national right to the city legislation in Brazil (Fernandes 2007) and the Occupy Wall Street protests.

In order to understand the basic practices of sociospatial secretion, it may be best to turn to a more extended delineation of a "real world" example in order to illuminate this. So instead of a six-storey house, consider the complex assemblage of Nigerian oil.

Spatial Secretion: The Nigerian Case

Whether a military or a civilian government has been in charge, a reign of "petro-violence" and oil-stoked corruption has been continual since the initial "black gold" strike in 1958. Despite some seven hundred billion dollars worth of oil extracted during the last fifty-odd years, per capita income is less today than it was in 1958. In the Niger River Delta itself, with its "606 oilfields," which "supplies 40% of all the crude the United States imports and is the world capital of oil pollution . . . , life expectancy in its rural communities, half of which have no access to clean water, has fallen to little more than 40 years over the past two generations" (Vidal 2010, 1). In his working paper of 2009, Michael Watts reports that the situation has only deteriorated: "A half century of oil wealth . . . has propelled Nigeria into the ranks of the oil rich at the same time as much of the petro-wealth has been squandered, stolen and channeled to largely political, as opposed to productive, ends. . . . 85 percent of oil revenues accrue to 1 percent of the population. . . . Around $300 billion of the $700 or so billion in oil revenues accrued since 1960 have simply 'gone missing'"(2009, 2). In terms of petro-violence, Watts reported in 2004 that "since the 1990s, there has been a very substantial escalation of violence across the [Niger River] delta oil fields, accompanied by major attacks on oil facilities" (2004, 7). In his working paper of 2009, Watts reports that the situation has only deteriorated: "According to a report released in late 2008—prepared by a 43 person government commission and entitled *The Report of the Technical Committee of the Niger Delta*—in the first nine months of 2008 the Nigerian government lost a staggering $23.7 billion in oil revenues due to militant attacks and sabotage" (2009, 2). To fill out this thumbnail sketch, the existence of the Nigerian resistance movement must be inserted, a movement that has arguably embodied the best and the worst of such movements, the best by the resistance of the Ogoni tribe, whose "influential and charismatic leader Ken Saro-Wiwa" was executed in November 1995 (Watts 2009, 2) and the worst by warlords such as Ateke Tom, leader of the Niger Delta Vigilantes, which is, according to Watts, nothing but "a vehicle for political thuggery and organized crime rather than any political project" (Watts 2009, 11–12).

Whatever its merits, a certain portion of the resistance movement in the Niger Delta has certainly become a force to be reckoned with, as well as a force that, at least to a certain degree, can claim to legitimately represent the will of the people. What seems to have happened in Nigeria is that "the shift from non-violent protest to militancy, and ultimately to armed struggle, was in many respects the inevitable result of the Nigerian government's brutal repression of the Ogoni movement" and the murder of Saro-Wiwa (Watts 2009, 2). And there seems to be no abatement to this violence, as in May 2009 "federal troops launched a full-scale counter-insurgency" against the insurgents, to which MEND (Movement for the Emancipation of the Niger Delta) responded in July 2009 with "an audacious, and devastating assault attack on Atlas Cove, a major oil facility in Lagos. . . . By the end of 2009," Shell (the primary oil extraction company operating in Nigeria) had "closed its western operations completely," abandoning its Delta facilities while keeping its "barely producing eastern facilities running" (Watts 2009, 4).

Now that I have sketched out this admittedly rudimentary history of oil production in Nigeria, I want to turn to the primary focus of this chapter, which is the organization of the Nigerian oil complex itself. Through this examination, I am able to trace the secretion of space in a mode that hopefully reflects Lefebvre's notions about the same.

The Oil Complex in Nigeria

In the draft of his 2009 working paper, Watts hurls a huge net around what he calls the oil complex, seemingly drawing everything into its encircling reach.

In seeing oil as a complex (rather than a production network) I want to emphasize the variety of actors, agents and processes, that gave shape to our version of carbon capitalism: that is obviously the IOCs [international oil corporations], the NOCs [national oil companies] and the service companies and the massive oil infrastructure, but also the petrostates, the massive engineering companies and financial groups, the shadow economies, theft, money laundering, drugs, organized crime, the rafts of NGO's (human rights organizations, monitoring agencies, corporate social responsibility groups, voluntary regulatory agencies), the research institutes and lobbying groups, the landscape of oil consumption (from SUV's of [sic] pharmaceuticals), and not least the oil communities, the military and paramilitary groups, and the social movements, which surround the operations of, and shape the functioning of the oil industry narrowly construed. (Watts 2009, 8–9)

Imagine that: this is the oil complex, *narrowly construed*! Broadly construed, we would have to include the environment and its "constituents," including such proximate members as the flora and fauna of the Niger Delta and such distal members as the waters of the world's oceans, and the atmosphere of the Earth. Let us also insert "the wildcat drillers, smooth-talking promoters, and domineering entrepreneurs" as well as the oil sheiks, Niger River gue-rillas, and CIA operatives running rigging companies as "company" fronts, and we only begin to scratch the surface of the motley crew populating the assemblage of characters circumambulating the vast territory of the oil complex (Yergin 1991, 13). These expansive construals of oil, though sweeping in scope and gigantic in scale, still fail to adequately capture the range and extent of this wide and rambling sector.

Perhaps if one slows down the predilection for cataloguing and proceeds incrementally instead, one may be better able to corral the vast undulating circle of this particular commodity chain. One might want to reference Stephen J. Collier and Aihwa Ong's construal of an assemblage as defining "new material, collective, and discursive relationships . . . as global forms are articulated in specific situations—or territorialized in *assemblages*. . . . These are domains in which the forms and values of individual and collective existence are problematized or at stake, in the sense that they are subject to technological, political and ethical refection and intervention" (2005, 4; italics, Collier and Ong). Perhaps Jane Bennett provides the most thoroughly latitudinarian definition of "assemblage":

> An assemblage is, first, an ad hoc grouping, a collectivity whose origins are historical and circumstantial, though its contingent status says nothing about its efficacy, which can be quite strong. An assemblage is, second, a living, throbbing grouping whose coherence coexists with energies and countercultures that exceed and confound it. An assemblage is, third, a web with an uneven topography: some of the points at which the trajectories of actants cross each other are more heavily trafficked than others, and thus power is not equally distributed across the assemblage. An assemblage is, fourth, not governed by a central power: no one member has sufficient competence to fully determine the consequences of the activities of the assemblage. An assemblage, finally, is made up of many types of actants: humans and nonhumans; animals, vegetables, and minerals; nature, culture, and technology. (2005, 445)

In the case of the oil assemblage, from a gas pump in Sligo, Ireland (for instance), to the speed boat of a squadron of bunkering MEND rebels swooping through the Niger River Delta to the corporate boardrooms of Shell Oil to various NGO offices in Lagos—well, to paraphrase Bob Dylan, that's a lot of territory to cover, indeed. What's even more is that we need to include a diachronic formulation of the assemblage as well, for "assemblages are in constant variation, are themselves subject to transformation" (Deleuze and Guattari 1987, 85). For nothing stands still, sitting patiently while philosophers and historians compose their analyses.

So, should one proceed synchronically or diachronically, does one lead with space or time, geography or history? Didn't I suggest, or rather *demand*, that such a bifurcation is exactly the thing to avoid whenever analyzing basically *anything*? Let alone when analyzing something so alive to the nuances of space and time as oil, crude itself a product of the passage of time in certain vectors of space, pockets sunk deep into the earth, the bones of dinosaurs transformed into the liquefied mush that powers our age.

In *Vibrant Matter: A Political Ecology of Things*, Jane Bennett, in this instance following Deleuze and Guattari's construal of assemblage from *A Thousand Plateaus* and paraphrasing herself to a certain degree, writes:

> Assemblages are living, throbbing, confederations that are able to function despite the persistent presence of energies that confound them from within. They have uneven topographies, because some of the points at which the various affects and bodies cross paths are more heavily trafficked than others, and so power is not distributed equally across its surface. Assemblages are not governed by any central head: no one materiality or type of material has sufficient competence to determine consistently the trajectory or impact of the group. (2010, 23–24)

This helps, especially in terms of the oil complex in its Nigerian manifestation, due to the fact that the oil assemblage as a whole has as many heads as an oily Medusa, what with MEND forces, multinational firms, and the various branches of the Nigerian government all struggling for dominance over the assemblage.

Outlining "some general conclusions on the nature of Assemblages," Deleuze and Guattari state:

> On a first, horizontal, axis, an assemblage comprises two segments, one of content, the other of expression. On the one hand it is a *machinic assemblage* of bodies, of actions and passions, an intermingling of bodies reacting to one another; on the other hand, it is a *collective assemblage of enunciation*, of acts and statements, of incorporeal transformations attributed to bodies. Then on a vertical axis, the assemblage has both *territorial sides*, or reterritorialized sides, which stabilize it, and *cutting edges of deterritorialization*, which carry it away. (1987, 88; italics, Deleuze and Guattari)

Without bothering to completely unpack this quite complex statement, one can garner the following from it in terms of the Nigerian oil complex. Collectively, it is an assemblage of enunciation, albeit one with many competing voices speaking in many disparate tongues, uttering the languages of bullets, contracts, environmental catastrophes, revolutionary polemics, tribal ties, resource extraction, drill bits, and governmental corruption, to name just a few of its elements. Simultaneously, it is also an assemblage that has both sides of the territorial coin referenced by Deleuze and Guattari, that which stabilizes and that which cuts away, though one would have to concede that in Nigeria the side of deterritorialization and fragmentation has been in the ascendancy throughout much of the period since oil was discovered in the Delta.

To fully assemble the components of the Nigerian assemblage, I cite a Willbros brochure dated 2000, delineating that company's work on the oil depot facilities at the Atlas Cove Jetty in Lagos. Before outlining the nuts and bolts of the operation, the brochure describes the purpose of the facility and its immediate history, which rationalizes their job: "The Atlas Cove Jetty in the Lagos Harbor area is used to off-load coastal tankers and pump petroleum products to the Atlas Cove Depot for storage. In March 1998, a vessel collided with the jetty, causing extensive damage and forcing operations to continue under hazardous conditions" (Willbros 2000, 1). The copywriters then insert what may justifiably be termed a metaphysical/prescriptive statement of the functional and normative purpose of the jetty: "Maintaining an uninterrupted supply to the depot is of paramount importance" (Willbros 2000, 1) before launching into an incredibly arcane account of what they are actually doing there:

> To ensure a secure and reliable supply, Willbros and Bilfinger+Berger Gas and Oil Services Nigeria Ltd. were awarded [a] contract for the installation of a single point mooring facility and the construction of a pipeline to the terminal. This fast track project involved a 12-meter diameter Catenary Anchor Leg Mooring (CALM) buoy installed at a water depth of 17.2 meters, utilizing an anchoring system consisting of six special-purpose marine drag anchors. The pumping system of product tankers, up to 50,000 DWT, is connected to the CALM buoy with 160 meters of 16-inch floating hose. The buoy piping connects to a pipeline ending manifold (PLEM) via 37 meters of 12-inch hose. A 3.9-km, 20-inch subsea pipeline extends from the PLEM to a directionally drilled shore approach. The 20-inch line then continues an additional 2.6 km onshore to the terminal, where it connects to a splitting manifold. Gasoline, kerosene and diesel are transported to the terminal where the serial batch interface is sent to one of two new 175 m3

interface tanks and re-injection facilities. The integrated system is controlled by voice and data radio communications between the offshore facilities and the on-shore control room. Services further included the installation of a SCADA system and telecommunications facilities, the development of O&M manuals, perfor-mance testing, operator training, commissioning and the provision of spare parts for three years operations. (Willbros 2000, 1–2)

With bullet-pointed items, Willbros then lets us know they have also per-formed four sets of studies to verify, determine, identify, reduce (or elimi-nate), and calculate various things about the operation:

- Hydraulic analysis to verify pipeline size and capacity for each product
- Surge pressure analysis to determine the effect of valve closures within the system
- Process hazards analysis to identify, and reduce or eliminate potential hazards
- Products interface handling analysis to calculate the length of contamination spread between batches (Willbros 2000, 1)

This is all well and good, with the engineering specs inscribed in an argot technical to the nth degree, and the verified, determined, identified, simpli-fied, and calculated analyses seeming to simultaneously ensure that it is a fa-cility permanent, reliable, packed, and calibrated with quantifiable technical know-how.

However, "late in the night of July 12th 2009, 15 MEND gunboats," packed with irregulars, "launched an audacious, and devastating assault on" the At-las Cove oil depot and blew it all to hell (Watts 2009, 4), most likely com-pletely obliterating the work Willsbros and its employees had performed in 2000, shattering the CALM buoy, liquidating the PELM manifold, extermi-nating the SCADA system, and, quite possibly, destroying the six special-purpose marine drag anchors, the serial batch interface, the two 175 m3 in-terface tanks, and the entire re-injection facilities as well. The raid was the culmination of a bloody summer full of operations by both government and insurgent forces. In May 2009, "federal troops" had "launched a full-scale counterinsurgency against what the government sees as violent organized criminals who have crippled the oil and gas industry. "Thousands of dirt-poor villagers in the region around Gbaramatu, southwest of the oil city of Warri in Delta State—an area known to harbor a number of militant en-campments including the notorious Camp 5—have been displaced and hun-dreds of innocent villagers killed. The causalities are almost wholly Ijaw, an ethnic minority who inhabit the creeks and lowland riverine environments where the Niger river empties into the Atlantic" (Watts 2009, 4).

Here we need to insert into the oil complex assemblage explosive parameters and ballistic trajectories of those bullets as they penetrated the cutaneous and subcutaneous integuments—the skins—of the Ijaw, their dripping blood and shattered bones, the graves of the villagers, their decomposing bodies, the tears of their survivors, plus, as mentioned previously, the flora and fauna native to the Delta area, including "vast mangrove forests and a wide variety of animals that are specifically native to the particular ecosystem prevalent in that area," much of which has been extinguished due to the extreme pollution levels in the Delta (Roth 2011, 1). "Because the oil makes its way into the water system of the Delta, drinking water becomes polluted and fish die. This obviously has negative implications for the local population of the Delta, which mainly live off fishery and farming" (Nigerian Students for Environmental Action 2010, 1). Into this very *un*-everyday concatenation of incidents in the Niger River Delta insert the very everyday scene of, say, a Honda Accord driver pulling up to a pump along Massachusetts's famous 128 corridor or, say, a Jeep Cherokee getting its gas tank filled along Highway 395 in the Owens River Valley between the Eastern Sierras and the White Mountain Range. These drivers and vehicles and gas stations and gas tanks and pumps and highways are, of course, also part of this assemblage.

Now let me return to the Delta and the events of 2009.

> Overall the oil and gas industry, on and off-shore, has been crippled. Shell has closed its western operations completely, and the eastern region is barely producing 100,000 b/d [barrels per day]. Many of the engineering, construction and oil service companies have withdrawn core personnel and in some cases withdrawn completely.... Hostage taking—not only of oil workers, but also politicians, even children—has become a major growth industry. In the industry parlance, the international oil companies no longer have a license to operate. (Watts 2009, 4)

So here must be added a few other items to this ever-metastasizing complex of Nigerian oil. What about kidnapping vis-à-vis personnel decisions, and what about industry parlance vis-à-vis licenses by which it is permissible to operate? What about the language in those licenses? What about the personal lives of the personnel arrayed across the Nigerian oil complex, from Shell oil execs to kidnapped children to Ijaw tribal chieftains? What is too much to insert and what too little? What is too insignificant and what too ephemeral? Should we insert the tonal inflections of voices on anonymous telephones as ransoms are demanded for kidnapped children? Should we include the movements of construction workers as they tighten lug nuts on the tires of company trucks? Should we include the handwriting of Michael Watts or the royalty payments for Yergin's best-seller, *The Prize*? Questions of parameters and limits plague this discourse, scattering it to the Nigerian winds as they make their course through the Delta and skim over the surface of the Atlantic seas.

Now that I have reemphasized Watts's concerns by citing them and thereby inserting them into the oil assemblage of the Niger River oil complex, let me insert some other things as well. What about the stink of that oil, the percolating stew of that crude as it sinks slowly into the Delta soil, the tactility of oil as it clings to feathers, dirt, and hands; the way in which numbers fail to work as tools by which to comprehend the extent of these compounded spills; Shell execs as they do their utmost to defer responsibility for these spills to the effects of bunkering and sabotage; the thrill of guerilla tactics and the joy of plain old robbery (for why should that thrill and that joy be beyond the scope of our assembled assemblage?); the uselessness of this present attempt to circumscribe such a massive assemblage; oil itself, its unrecognizable matter-of-factness, its state of simply being there, innocent, dumb, resourceful in its very everyday unawareness of its great majesty as a resource; Shell headquarters with its corporate boardrooms, parking spaces, security, computers, secretaries, janitors, and so on; the energy departments of various governments across the world, all factoring in the amounts of oil needed for their respective citizens, and so on?

Secretion of the Oil Complex

First, let me bring spatial secretion back into the frame. In *The Production of Space*, Lefebvre states that "the spatial practice of a society *secretes* that society's space; it propounds and presupposes it . . . it produces it slowly and surely as it masters and appropriates it" (1991; italics, mine). What is interesting in the Nigerian case (and, I suspect, in many other cases as well) is that only the first portion of Lefebvre's dicta holds true; while the secretion of spatial forms in the Nigerian oil complex does *proceed* from the presuppositions and demands of a society dependent on oil for either its revenue or its energy, it does not follow that these spatial forms, once secreted, are *mastered* by the very society that secreted them. Neither of these forms, whether they be the oil itself or the structures connecting its extraction to its consumption, can be said to be mastered by the society which appropriates it, as oil is routinely bunkered, and structures as well as equipment are blown to bits.

Lefebvre also ties spatial secretion to the production of merchandise: "Space, which seems homogeneous, which seems to be completely objective in its pure form, such as we ascertain it, is a social product. The production of space can be likened to the production of any given particular type of merchandise . . . . there are interrelationships between the productions of goods and that of space. The latter accrues to private groups who appropriated the space in order to manage and exploit it" (Lefebvre 1976, 31). Consider Lefebvre's claim that space can be likened to the production of merchandise, and try to ascertain if it applies to the Nigerian oil complex. Well, a special type of merchandise, oil, being required by both developing and developed nations, and its extraction, in turn, requiring other special forms of merchandise such as tankers, oil depots, pipelines, rigs, and so on and so forth—all of this produces spatial formations, does it not? Here, the type of merchandise creates certain forms of spatial production, certain buildings, and certain reconfigurations of the Earth, its spatial forms, its environment. It might be simpler to stipulate that space is transformed due to the construction, extraction, and transportation requirements of the oil business. So, in the Nigerian case, for instance, during the 1970s, the state, "flush with money" from its share of the oil revenues, "through its fiscal linkages—road building, electrification, infrastructure development, industrial promotion, and so on—entered upon a massive industrialization program including autos, iron and steel and so on" (Watts 2000, 22). Here, we can extend the secretion and sociospatial production of space from the oil assemblage to include that which the revenue from oil extraction funds: roads, universities, as well as mansions, armed guards, American-made missiles, and so on.

The other reason that the production of space can be likened to the production of merchandise is that we are living under a regime of capitalism. In such a regime, space is defined using the marker of exchange, not use, value. Therefore, space becomes abstracted into a monetary value just as does any piece of merchandise under such a regime. So many coats will buy you so many acres of land, in other words. And under a strict private property regime, such as in the United States, the protection of space in its form of exchange value typically receives a higher priority than the use value that may be derived from that same space.

In any account of the practices of spatial production, formal as well as informal outcomes must be included: assessing the Brazilian housing situation without dealing with the *favelas* would be ludicrous; investigations of Wall Street with only an eye on the de jure trades without the other eye tabulating up the de facto trades would also be an exercise in futility. This is just as true when dealing with the oil complex in Nigeria, as the "monitoring agency Transparency International typically ranks Nigeria as one of *the* most corrupt countries in the world. . . . It is not without good reason that the National Electric Power Authority, or NEPA, was (and is) popularly assumed to stand for 'Never Expect Power Again'" (Watts 2000, 24; italics, Watts).

Here the Nigerian electricity-generating plants must be subsumed within the oil assemblage, as the secretion of electricity and the buildings and spaces through which electricity is produced as well as the substations and lines through which that electricity is meant to be distributed—in fact, the entire Nigerian power assemblage—has been severely impacted by the oil regime. Corruption spreads the secretion out into every aspect of society, to such a degree that "the depth of graft and bribery" caused "one disillusioned Nigerian commentator" to remark that "'Nigeria is not a country, it is a profession'" (Watts 2000, 24).

The production of space included within the Nigerian oil assemblage must also include the secretion of the assemblage as it manifests itself in London, the capital of its former metropole. Massey brings this to our attention in *World City*, in which she states that "oil and gas account in one way or another for about a quarter of London's stock exchange; Shell and BP have major offices and headquarters in London; London [itself] is utterly dependent on oil" (2007, 201). Massey also informs her readers of a number of London-based activist campaigns, including "the radical London collective, PLATFORM" and "London Rising Tide, a group campaigning around the root-causes of climate change," with the latter group making it their business to make explicit the ties between the UK and the Nigerian oil complex (2007, 203, 202–3). Massey also references Carbon Web Newsletter #2, which presents a map demonstrating that "if all the sources and links in the [Nigerian] oil commodity chain and its multifarious supports were mapped, the centre of London would be crowded with references" (2007, 205). So one can trace the Nigerian oil complex to the nexus of London and code that nexus with links leading back to the crude being extracted in the delta.

In fact, if one is going to be exhaustive in the excavation of the Nigerian oil assemblage, the entire historical relationship of the core/metropole, England, to its periphery/protectorate, Nigeria, should be included. This would transpose the complex into the component of time as well as that of space.

This seems to be the most realistic as well as the most veridical way to go, as what time has etched and embedded into the contours of both nations cannot be divorced from the contemporary actuality on the ground. Oil oozing out of Nigeria right now, being extracted, processed, and transported for use in the global North, cannot somehow be de-linked from a past that has led to its very extraction. So the entire colonial and postcolonial mode of the exploitation of resources needs to be inserted within the extended frame of this assemblage.

We should also insert insurgent commanders such as the Ijaw chieftain, Government Ekpemupolo (aka Tompolo), who reportedly has played both sides of the Nigerian oil equation, leading MEND forces while also serving as a consultant to Shell and other oil companies. "He was clearly a double agent, and his methods were very widely mimicked among the more privileged sections of those engaged one way or the other in the Niger Delta struggle. . . . Through him and many other Ijaw leaders, governments and oil companies made a lot of concessions in order to buy a peaceful atmosphere to move crude oil into the world market" (Nnanna 2009, 1).

Experts and academics such as Michael Watts should be inserted into the cataloging of the oil assemblage as well. After all, Watts, serving as an example of this group, has spent the majority of his academic career tracking the Nigerian oil complex while writing or editing books such as *State, Oil, and Agriculture in Nigeria*, published in 1987, and *Curse of the Black Gold: 50 Years of Oil in the Niger Delta*, published in 2008. In a sense, Watts has tied his professional career to the oil complex and the unpacking of the Niger River assemblage. The Niger Delta: Economies of Violence Project, with its twenty-five working papers, most either authored or coauthored by Watts, is cosponsored by the United States Institute of Peace, based in Washington, D.C.; Our Niger Delta, based in Port Harcourt, Nigeria; and the Institute of International Studies, based at the University of California, Berkeley, where Watts is a professor in the geography department, thus implicating all of these various organizations and institutions at least into the outer strands of the great web of the Nigerian oil assemblage.

Here what I want to underline is that every one of the branches of this assemblage must secrete/produce its own spatial node within that assemblage, whether that space be a gas station built on the edge of an Arizona subdivision, an oil subsidiary only existing on paper for the purposes of tax evasion, an insurgent encampment in the far reaches of the delta of the Niger River, or a professor's office on the campus of the University of California, Berkeley.

Now that the Nigerian oil assemblage has nearly been extended to the snapping point, let me return to our primary subject, the everyday.

People are driving cars, an everyday action occurring in space. They are filling up gas tanks, again an everyday action occurring in space. Some people are working on oil crews, actions that are more or less everyday as far as such workers are concerned. Others are bunkering oil or toting firearms as they cruise through the Niger Delta, again more or less everyday actions for them. Others are making decisions or pushing paper around in Shell and BP headquarters in Europe, more everyday acts for those involved in such acts. My point here is that far-reaching assemblages occurring in space, such as the oil complex in Nigeria, no matter how expansively construed or how "exotic" the actions constituting them may seem, always track back to the everyday and the more or less contained actions of respective quotidian domains.

One final point. The Nigerian oil complex is, of course, only one assemblage among an almost numberless variety that may have been chosen as the prime exemplar of spatial secretion and its connection to the everyday. For instance, the international market in organs, as elucidated by Nancy Scheper-Hughes in "The Last Commodity: Post-Human Ethics and the Global Traffic in 'Fresh' Organs," with its ties from the University of Maryland Medical Center, whose "website advertises its kidney transplant program in Arabic, Chinese, Hebrew, and Japanese" (2005, 150), to the "rustic little village of Mingir, Moldova," and the "slums and shantytowns half a world a way [*sic*] in the Filipino capital of Manila" (2005, 152) where young men and women sell their kidneys to organ brokers, to the bioethicists attempting to impose some sort of deontological order on this regime, to the "highly fetishized kidney" itself, invested with all the magical energy and potency that the transplant patient is looking for in the name of 'new' life" (2005, 156), is another exemplar that could easily fit the bill. So is the telegraph system of the 1800s, as elucidated by Andrew Barry in "Line of Communication and Spaces of Rule," with the telegraph's ties to the development of the standardization of electrical measurement and its role as an early warning system for the British Empire, as well is its "function as a *mediating machine*, articulating a relation between the most abstract and fundamental areas of physical science and the most practical and urgent problems facing the empire" (1996, 134; italics, Barry) and its place as a pivot point in scientific methodology, as the "disastrous failure" of the first attempt at laying down a trans-Atlantic telegraph cable served as a proverbial wake-up call for scientists, engineers, and politicians that the arcs of ambition and technology could not be aligned unless "adequate" and "appropriate" standards as well as the instruments by which to calibrate such standards were in place (1996, 133). The telegraph assemblage also serves nicely as a prototype for many of the questions in regard to the technological capacities of communication and information systems: how to balance the positive and negative forces of, respectively, the distributive and emancipatory reach of these systems with their potential as repressive surveillance tools.

Linking these two examples to spatial production is not difficult, of course. The trans-Atlantic cable is a wonderful example of a feat of spatial secretion, as are the secret locations in the Philippines, Turkey, Brazil, and Moldova where illicit organs are surgically removed to be transported to the hospital operating rooms of Israel and the United States, where those organs are implanted into patients at a cost of upward to $200,000. The latter offers up an example of what Lefebvre might have called the invisibility of their creation. "The fact remains . . . that productive [spatial] operations tend in the main to cover their tracks; some even have this as their prime goal: polishing, staining, facing, plastering, and so on. When construction is completed, the scaffolding is taken down" (1991, 113). This tendency to cover the tracks of that which is created obfuscates the labor power that has been necessary for creation and production, and it also makes it much more difficult to trace the inner workings of the everyday, as a veneer is placed between consumption and production precisely to cover the scaffolding necessary for consumption to proceed. Hence, in their advertising, Shell, Exxon, and BP present oil and gas as commodities divested of extractionary operations and all that those operations may entail, including environmental destruction, violence, and the possible impending collapse of the global ecosystem, which, if it occurs (and all signs point to this possibility becoming a reality), will be primarily caused by the oil assemblage, most especially our seeming incapacity and unwillingness to unhinge ourselves from its grip.

Conclusion

Through a study of the ruse as theorized by de Certeau and as practiced by the fictional characters Svejk and Schweyk as well as their creators, Hasek and Brecht, I have tried to shed light on both a primary tactic of everyday existence—the ploy or the ruse—as well as to bring history into the fold of this schema. Through spatial secretion as theorized by Lefebvre and as practiced in the oil fields of Nigeria, I have tried to shed light on the spatial ramifications of everyday existence. Spatial secretion throughout multiple assemblages is one of the great constituents of the everyday. Without it, we would be without space and thus nowhere. Next, we turn to Marx and an examination of another primary factor of every existence, reproduction.

CHAPTER 4

What Marx Brought in
from the Cold Reproduction

"Labour-power exists only as a capacity of the living individual," writes Marx in volume 1 of *Capital*. "Its production consequently presupposes his existence," he adds. "Given the existence of the individual, *the production of labour-power consists in his reproduction of himself or his maintenance*" (1994, 268; italics, mine). Or as Giddens puts it: "For Marx, history is a process of the continuous creation, satisfaction and re-creation of human needs" (1971, 22).

Self-maintenance, or what could be called the reproduction of the self in order to survive yet another day, is one of the primary components of everyday life. Eating and drinking, sleeping and the elimination of waste matter—these constitute the bare minimum of one's circadian reproductive maintenance; that is, unless the laborer's value has diminished so much that one's reproduction is deemed unnecessary, as has often been the case with slaves as well as with industrial workers in certain periods. In such instances, production without any subsequent reproduction is judged to be sufficient: the death of the laborer is accepted while the elemental principle of maintenance is nullified. The extent to which basic needs are met is contingent: "His natural needs, such as food, clothing, fuel and housing vary according to the climatic and other physical characteristics of his country" (Marx 1994, 268). Yet these factors are also conditioned by "the habits and expectations" with which "the class of free workers has been formed," placing the value workers put on their own labor within the compass of "the level of civilization attained by a country," instead of somehow being exogenous to it (Marx 1994, 268). Indeed, one of Marx's greatest accomplishments is his insistence that the working class be included as a certified component of "civilization" and therefore worthy of investigation, thought, and even, at least in Marx's case, valorization.

What I would like to accomplish in this chapter is to add the processes of reproduction to the tools of analysis by which to circumnavigate the everyday. The "habits and expectations" of reproduction are crucial components of the everyday world and thus worthy of both investigation and inclusion within this study, a conclusion that may seem obvious, and even obvious to a banal degree, but the ramifications of which may prove to be of great interest.

However, before delving into this, it should also be acknowledged that it is not simply the people of whatever class that must be reproduced on a day-to-day basis, but the entire society and the built environment as well. These must be reproduced on a temporal and spatial basis that must be more or less aligned with the flows and the needs of capital (at least in society as presently constituted). Such requirements are necessary to maintain the fluid mobility of easy transportation and well-orchestrated logistics, as well as the long-term needs of capital for the health, the education, and the discipline of the workforce.

Marx tells us in *Capital* that reproduction is a ceaseless process that cannot come to cessation in any society, no matter what the mode of production may be. And David Harvey points out in *The Limits to Capital* that in capitalistic societies entire social classes have to be reproduced as well. But Harvey also notes that reproduction must proceed on an individual biophysical basis as well. "The metabolic processes which permit ... internal *self-reconstitution* to proceed entail exchanges with my environment and a whole range of trans-formative processes which are necessary for the maintenance of my bodily individuality" (Harvey 1995, 6; italics, mine).

Marx ties together production and reproduction by pointing out that workers must produce enough surplus value for the capitalist to create a profit and the variable capital (wages) for their own reproduction. Without such an arrangement, workers will not be employed and will have to either find other means for their own reproduction or die.

Furthermore, roads, railroad lines, bridges, schools, and, indeed, the entire built environment, once constructed, is also subject to reproduction. Typically, the capitalist, being a capitalist, attempts to download as much of the costs of this reproduction as possible on to the state (the public), while reserving the right to collect a maximum of the benefits accruing to such communal outlays. "Individual capitalists find it hard to make such investments as individuals, no matter how desirable they may regard them" (Harvey 1985, 8). Capitalists must also "as a class ... if they are to reproduce themselves, continuously expand the basis of profit" (Harvey 1985, 1), meaning that they must continuously produce as well as reproduce buyers for their goods.

But besides continuously expanding the basis of profit, capitalists must also reproduce themselves, both individually and as a class. The cyclical patterns of capitalism tend to reproduce themselves as well. Robert Brenner outlines precisely such a financial-economic self-perpetuating cycle in *The Economics of Global Turbulence*: "As a consequence of the continuous, precipitous fall in profit rates that resulted from the worsening of global over-capacity and intensifying international competition between the later 1960s and early 1980s, there emerged, in classical fashion, a dual problem of weakening aggregate demand and weakening productivity growth, which tended to be *self-perpetuating*" (2006, 280; italics, mine).

Reproduction at a greater and greater scale merely in order to "maintain position" has been a necessity from the very inception of capitalism. "In marked contrast to feudalism and earlier rank-redistributive and primitive subsistence economies, merchant capitalism was a self-propelling growth system in which the continued expansion of trade was vital" (Knox, Agnew, and McCarthy 2008, 103). But such a necessity created its own contradictions:

> Mercantile success required the merchants to buy as cheaply as possible, and to sell as expensively as possible; it also demanded that they trade in as large a volume of goods as possible. . . . This created a contradiction, however, for the producers were also consumers (though not of the goods they produced), so that if the prices they received were low, they could not afford to buy large quantities of other goods and thus satisfy the demands of the merchant class as a whole. A consequence of this was a great pressure on producers to increase the volume of goods for sale, which meant increasing their productivity, while merchants put pressure on consumers to buy more, even if this meant borrowing money in order to afford their purchases. Both processes . . . involved producers raising loans which they had to repay with interest [thus ensuring the reproduction of the financiers]; to achieve the latter, they had to produce more (or, if they were employees rather than independent workers, to work harder). (Johnston 1980, 33–34, quoted in Knox, Agnew, and McCarthy 2008, 103)

From this can be gathered that reproductive loops, whether of a positive or a negative nature, have been an elemental part of capitalism from its inception.

It should also be added that the three circuits of capital—commodity capital, money capital, and productive capital—are continually reproducing themselves as they are transformed into one another. As Marx puts it: "One part of capital, continually changing, continually reproduced, exists as a commodity capital which is converted into money; another as money capital which is converted into productive capital; and a third as productive capital which is transformed into commodity capital" (1997, 109). So the entire loop that capital makes is equivalent to all three phases as they operate together:

capital, production, and commodification in a continuous reproductive cycle, tied together by their capacity to reproduce one another.

Reproduction, economic cycles, and the like can also be linked, to a certain degree at least, to what can be termed the efficacy of multiplier effects. Is it too poetic to stipulate that reproduction is simply multiplier effects multiplied on a more or less continual basis? Here is Harvey on the *effects* of multiplier effects:

> Investment in the built environment has a significance far beyond the direct investment that it absorbs. First, these investments generate certain multiplier effects because the subsequent use (and hence the value) of an urban infrastructure depends upon the commitment of further resources. Under private property relationships and a capitalist mode of consumption, these multiplier effects can be quite substantial (most households are heavily over-equipped relative to their needs). The construction of housing, for example, requires complementary investments in transportation, household equipment, community facilities, and the like, while it also imposes a variety of recurrent operating costs (energy inputs being a prime example). . . . In addition, construction activity has certain multiplier effects in the employment sphere. (Harvey 1974, 4)

So this might be termed a virtuous circle of the capitalistic variety, wherein housing construction catalyzes transportation outlays, expenditures on durables (cars, refrigerators, washing and drying machines, etc.), construction of Little League parks and senior centers, and so on. Of course, housing developments can go south as well. For instance, by the willful neglect of arranging for such effects to multiply, quite a number of housing tracts, particularly in the Southwest of the United States, have been built in the middle of nowhere without logistical networks in place. A hapless buyer may purchase a home with no connecting roadways to "the outer world," no schools, no linkages to water systems, and no hookup to the electric grid and the gas network. Caveat emptor, indeed. I expand on some, if not all, of these points throughout the course of this chapter. I simply want to indicate the vast scope of reproduction in a capitalistic society.

I proceed down three disparate avenues, one for the working class, one for the middle class, and one for the ruling class. In the first case I demonstrate how members of the working class in China are reproducing themselves in a manner that defies the stereotypical expectations of both communism and capitalism. In the second case I show how one branch of the middle class, that of the United States of America, is teetering on the brink of being unable to reproduce itself. And in the final case, I display the excesses upon which or through which the global ruling class is reproducing itself, or, perhaps more accurately, *over*reproducing itself. And though I may neglect to mention the everyday as often as I might (or as I should), bear in mind that all these delineations of groups and classes can be distilled down to the everyday actions of people working, sleeping, dwelling, eating, defecating, drinking, urinating, and so on.

The Surplus Army of Labor and the Working Class

"The flight of serfs into the towns continued without interruption through the entire Middle Ages," writes Marx in *The German Ideology*. "They never managed to organize themselves and remained unorganized rabble. The need for day labor in the towns created the rabble" (Marx 1994, 134). However, "the rabble" ended up fulfilling much more of a role in the history of feudalism and capitalism than merely being available for day labor, though they certainly fit that bill and at cheap prices as well. As members of what Marx calls the reserve or surplus army of workers, the rabble has also stood as markers or signals to "ordinary" workers that there is a group of desperate men and women willing and at least somewhat able to readily take their jobs, should ever that need arise. In fact, to fulfill the function of serving as a sign to workers that other things being equal, things certainly could be much worse, the rabble doesn't even have to be employable, they merely have to survive. Then they can exist as living symbols, human signposts, of what can happen to those who somehow slip from the halter and either cannot or will not work. Engels, writing of Manchester and Birmingham in the 1840s, describes this semiotic operation thusly: "While I was in England at least twenty or thirty people died of hunger under the most scandalous circumstances. . . . Of course, only a few [of the poor] actually die of hunger. But what guarantee has the worker that this will not be his fate to-morrow? Who will give him security of employment?" (1958, 32).

Moreover, the reserve army can be counted on whenever capitalists require (or simply desire) a quick influx of cheap labor, depressing wages as they flood into the workforce, a positive by-product of their recruitment as far as the capitalist is concerned. "The role of the Reserve Army is to provide a pool of available labor to satisfy sudden bursts of activity characterizing expansion phases of the business cycle. . . . That inadequate labor supply does *not* constitute a constraint upon the 'elasticity' of the system is accounted for precisely in terms of the operation of the Reserve Army" (Hollander 2008, 100; italics, Hollander). Giddens also notes the importance of the reserve army in Marx's schema as a necessary ingredient for the capital class to ensure themselves against economic stagnation as it "acts as a constant depression upon wages" (Giddens 1971, 56).

But then the question arises: how are these people to be maintained until that time? What are the means of their reproduction until they are needed? Frequently, this is the nexus in which capital welcomes government into the equation, as the state can be used to pick up the slack, offering up such benefits as food stamps and unemployment insurance to reproduce this sector until its members are required by capitalists. However, when the workforce becomes a global entity, then the requirement for the reserve army to be maintained and reproduced at a local, regional, or even national scale can become problematic or even unnecessary. Perhaps this juncture is when the reserve army becomes dispensable, only able to fulfill a role as a living cautionary tale on a massive transnational scale to those who may be entertaining the thought of refusing to abide by harsh austerity measures such as the slashing of wages and benefits.

From the earliest days of the capitalist mode of production, capitalists perceived that a barely sufficient wage created the most efficient means to keep workers alive while simultaneously ensuring that they were forced to return to work, day after day. As Bernard de Mandeville so sweetly articulates this operation in *The Fable of the Bees* (quoted by Marx in the first volume of *Capital*), such processes work thusly: "It would be easier, where property is well secured, to live without money than without [the] poor; for who would do the work? . . . As they [the poor] ought to be kept from starving, so they should receive nothing worth saving. . . . It is the interest of all rich nations, that the greatest part of the poor should almost never be idle, and yet continually spend what they get. . . . Those that get their living by their daily labour . . . have nothing to stir them up to be serviceable but their wants which it is prudence to relieve, but folly to cure" (Mandeville 1728, 323, 324, 328, qtd. in Marx 1996, 610).

Though this almost refreshing frankness on the part of Mandeville seems to be a quaint relic of the past, the sentiment has surely been revived. After decades of struggle led to some concessions on the part of capital and the state to guarantee the welfare of the poor and the working class, following the triumphalism of Thatcher and Reagan a reverse course has been set, as if Mandeville's standards had once again become the norm, and any socioeconomic progress that had been achieved was to be nullified by an ascendant upper-class elite. It should also be added that for Mandeville to claim that, say, his "free nation" of England is surviving and even thriving without recourse to slavery is a bit disingenuous, as English merchants and the nation as a whole were heavily invested in the slave trade.

The movement of European serfs into the towns during the Middle Ages and the subsequent formation of the "rabble" from which day laborers could be extracted serves as a harbinger of the eradication of English peasants from their small plots of land during the long series of enclosures stretching roughly from the late 1500s to the mid-1800s, with "the years between 1760 and 1820 . . . the years of wholesale enclosure" (Thompson 1963, 198). It is also a foreshadowing of a trend that Harvey reminds us is still occurring, as capital relentlessly seeks out cheap pools of labor. "All around the world the integration of rural and hitherto independent peasant populations into the workforces has occurred" (Harvey 2010, 58–59). A long wave of eradication and enclosure has encircled the globe during the last five to six hundred years, a wave that is still removing peasants and farmers from their land and forcing them into the industrial workforce at cut-rate wages. While it is true that this process may have started in Western Europe, particularly in England, it has metastasized since then to encompass nearly every corner of the globe.

This wasn't always accomplished through literal enclosure. For instance, in El Salvador "all communal lands were abolished" in 1882, leading "to a redistribution of about 40 percent of all agricultural land" (Nugent and Robinson 2010, 68).[1] This concentration and "control of land" by the elite "'created a large proletariat and semi-proletariat of agricultural wage laborers. . . . Through the use of force, squatting was held in check and the landless . . . became dependent on coffee growers for survival'" (R. G. Williams 1994, 124, qtd. in Nugent and Robinson 2010, 68–69). In this case, subsistence farmers were forced off the land to become coffee workers rather than factory laborers, as in the English case. However, the commonality between English peasants and their Salvadorian counterparts lies in the coercive removal from the land itself and the formation of a low-wage workforce openly available for exploitation. Moreover, in El Salvador the judicial system was recruited to keep workers on the plantations. "A system of 'agricultural judges' was created in 1881 to enforce restrictive vagrancy laws intended to impede labor mobility and trap workers on coffee estates" (Nugent and Robinson 2010, 69).

In the following passage, Raymond Williams describes the situation in England: "Much of the real history of city and country, within England itself, is from an early date a history of the extension of a dominant model of capitalist development to include other regions of the world" (1973, 279). So that what happened from 1760 to 1820 to the "old rural hand weavers with hearth and home" of England is now being replicated in, say, Vietnam, as peasants are pressed to leave the subsistence economies of their villages and enter into the wage-earning economies of the factory (Engels 1979, 26). When we also factor in the recruitment of women into the workforce, the global reserve army becomes a massive repository from which capital can draw when and where it will. Many of these people become employed at the lowest levels of the working classes in the factories of China, Southeast Asia, and Latin America. But others exist at the margins between starvation and reproduction, a surplus army of the unemployed, the underemployed, millions upon millions living in *favelas* and shanty towns, millions and millions "houseless," as Engels calls those without shelter in *The Condition of the Working Class in England* (1958).

So members of the reserve army cannot be unskilled and immobile, as to "properly" fulfill their role they must remain skilled (or perhaps even develop new skill sets), as well as be willing to relocate anywhere and at any time the mere possibility of employment beckons. The other factor depositing prospective workers into the repository of the urban reserve army are the magnets of familial attachments and personal relationships continually drawing rurally based relatives, friends, and neighbors into "the great towns" (Engels 1958, 30). As Farha Ghannam puts it, writing about migratory patterns in contemporary Egypt, there is a "flux of families from rural Egypt who move to Cairo to improve their living conditions; a network of connections draws relatives, neighbors and friends from the countryside" (2008, 275).

To retrieve a question posed above, how does this reserve army maintain itself until it is needed? True, governmental assistance abetted by odd jobs within the enormous sector of the informal economy can more or less reproduce the reserve army until its services are required by the capital class. But how can members of the surplus army retain their skills (given that they had skills to retain in the first place) if they are not employed?

In Albert Cossery's novel *The House of Certain Death*, set in the Egypt of the late 1940s, Ibrahim Chehata, an unemployed carpenter, demonstrates that one can work, or at least engage in the pretense of working, and still be a member of "the rabble," so-called. "The carpenter had set up business in a corner of the courtyard, since he was too poor to rent a shop. One could always see him working furtively at some small job, but this continuous labor only served as a pretext to conceal his tragic condition" (1949, 18).

However, as already stated, the entire question of the retention of a skill set by an individual worker or the maintenance of a reserve army on a national scale becomes nugatory if the workforce is accessed on a global scale. Then it becomes some other country's problem by a default of the pass-the-buck variety. Obviously, any sense of national obligation for the reproduction of surplus labor will be tenuous at best and nonexistent at worst if the reserve army is being recruited from outside the nation's borders. Any appeal to a sense of national shame is unavailable if the nation no longer contains within its boundaries the workers being exploited and the destitute being immiserated, that is, if the workers are laboring in another country. Many of the appeals concerning amelioration for the poor and the working class in England in the mid-nineteenth century are at least partially based on such a patriotic sentiment. Engels cites a Dr. Hawkins, who implicitly invokes a certain sense of national pride as he describes the lot of the "factory classes" of Manchester who fall far short of the "national standard" of typical English folk. "I have never been in any town in Great Britain nor in Europe in which degeneracy of form and colour from the national standard has been so obvious" (1958, 178).[2]

Apply a multiplier effect to the abandonment of national shame about the condition of the working class, the working poor, and the destitute with one country after another relinquishing whatever sense of responsibility it may have felt to be due to the marginalized, and the reproduction of the reserve army becomes no nation's job. Hence the globalization of responsibility into a virtual zone that turns out to be a void.

What has been witnessed in the last thirty to forty years, that is, from roughly 1973 to the present (2017), is that the working class and much of the middle class have been driven into the ranks of the reserve army, or at least they have been pressed into proximate alignment with those ranks. In *The Enigma of Capital*, Harvey cites Alan Budd, "Thatcher's chief economic officer," who confesses "that 'the 1980s policies of attacking inflation by squeezing the economy and public spending were a cover to bash the workers' and so create an 'industrial reserve army' which would undermine the power of labour and permit capitalists to make easy profits ever after" (2010, 15).[3] In *The Boom and the Bubble*, Brenner pushes this downward pressure on wages back to the 1970s: "Supported by increasingly sympathetic governments, employers across the advanced capitalist world unleashed an ever more aggressive attack on workers' organizations and workers' living standards" (2002, 24). This proved especially "successful" in the case of the United States, as "manufacturing real wages in Germany and Japan increased by about 35 per cent over the decade 1985–95," while "in the US they inched up little better than 1 per cent" (Brenner 2002, 61).

This engrossment of the reserve army, along with the concomitant depletion of the working class, has produced a situation in which the reproduction of both the surplus army and the working class has become a tenuous affair, whereby one out of five children in the United States is officially living in poverty, a state of affairs that doesn't seem to have caused any great outburst of national shame, or even the slightest quiver of patriotic alarm. Yet at the same time, potent forms of ideological persuasion—broadcast via a combination of the allurements of advertising, the "thinking" of the punditry, the exhortations of the political class, and the ravings of talk show hosts and bloggers—have at least partially convinced many members of the working class that there is no such thing as the working class and that even if there were such an entity, its members would have nothing in common with members of the surplus army.

But much of this is the fault of thinkers on the left as well, as they have steadfastly clung to a definition of the working class that is antiquated. While such a definition (basically equating the working class with factory and mine workers) may have made sense in the England of Marx and Engels's era or the Detroit of Henry Ford's epoch, it no longer makes any sense today and has long been in need of a complete overhaul.

For confirmation of this need to expand the definitional parameters of the working class, all that's needed is turn to Los Angeles and the campaign to organize maintenance and carwash workers, resulting in the Justice for Janitors and the Clean Carwash campaigns. It is safe to say that janitors cleaning up corporate headquarters and carwash workers wiping down Escalades were not the primary target of the call for working-class revolution issued by Marx and Engels in the middle of the nineteenth century; yet it is also hard to imagine Marx and Engels not agitating for just such campaigns, which target the "working poor," including health-care providers, maids, nannies, housekeepers, as well as the aforementioned janitors and carwash workers (Soja 2010, 111). Obviously, at least some members of the working class understand this, as it was the leadership and the rank-and-file of L.A.'s Local 399 of the Service Employee International Union who organized janitors and led them on a successful strike in 1990. Primarily undocumented immigrants from Latin America, the janitorial workers of Los Angeles "effectively symbolized the New Economy in Los Angeles" (Soja 2010, 140). Members of the United Steelworkers understand this as well, as they represent the carwash workers at Bonus Car Wash in Santa Monica, which in October of 2011 signed a labor agreement with its employees.

So in Los Angeles, the ranks of the working poor were used exactly as Marx predicted, not only in an absolute sense as a source for cheap labor, but also in a relative sense as a tool to discipline and immiserate other workers and knock them out of the ranks of the middle class or even of the lower middle class. That "the expanding army" hailed from deeply poverty-stricken regions of the globe only adds to their capacity to perform these functions. However, their immigrant status also aided in their predilection to unionize, as many members of the Los Angeles contingent of the surplus army arrived from countries rich with histories of union activity. "Traditions of militant labor organizing were particularly intense in the Korean community and among the nearly one million immigrants from El Salvador, Nicaragua, and other Central American countries" (Soja 2010, 140). In the highly successful Justice for Janitors campaign, the stereotypes of the docile apolitical immigrant and the superficial apathetic Angelino were turned on their heads, as militant workers drew support from almost every layer of the population of Los Angeles, from Beverly Hills dowagers to East L.A. busboys. What had been perceived as a highly manipulable sector of the populace was demanding an upgrade from the lowest ranks of the working poor. In the process, the janitors improved their own quality of reproduction, as rising wages and better working conditions ameliorated obstacles to their own reproduction.

But to return to the original question, how does the reserve army as well as the working class reproduce itself? Of course, such processes will vary widely, depending on where the lens is positioned, so let me survey only one variety of sociogeographical milieu in hopes of at least partially indicating the ways in which reproduction is occurring among the reserve army, and that is how it is manifested in contemporary China.

In the urban villages of China, the residents of rural villages that have been gradually "surrounded or encroached upon by urban expansion" are reproducing themselves through a means that runs counter to the story I have been telling about rural folk being absorbed into the urban workforce (Zheng et al. 2009, 426). As cities across China have expanded during the last twenty-five years or so, they have "acquired" villages, but the village collectives have maintained "ownership" of their land, making for an odd hash of cockamamy forms of capitalism and communism resistant to facile classification as either one or the other. "While farmland in villages on city fringes may have been acquired by city governments, land for housing continues to be owned by village collectives and is allocated to village residents. However, many urban village residents have given up farming and have instead built or expanded housing to rent to migrants. To these villagers, rent has replaced agriculture as the main source of income" (Zheng et al. 2009, 426). So while the village itself is owned by the circumambient city and, ultimately, by the state itself, the "land for housing" is owned by the residents of the village. And what do these villagers do? They abandon agriculture and become landlords. Well, this would surely be a puzzle to Marx and Engels. Not to mention Chairman Mao. And these are not isolated cases that have occurred in only one or two cities. Such cases are happening in nearly every major Chinese metropolis. Urban villages as well as urban villagers renting out their land as housing are common in China. In "Urban Villages in China: A 2008 Survey of Migrant Settlements in Beijing," Siqi Zheng, Fenjie Long, C. Cindy Fan, and Yizhen Gu inform us that "hundreds of urban villages exist in large cities such as Guangzhou and Shenzhen" (Zheng et al. 2009, 426).

For their study, Zheng et al. randomly selected 50 urban villages "from the total of 867" of such villages in the Beijing Metropolitan Area (2009, 428). In these villages, migrants outnumbered "natives" by 478,525 to 91,656, a ratio of approximately 5:2 (Zheng et al. 2009, 429). "Facilities in the urban village units are inadequate and poorly maintained. More than 90 percent of the surveyed units do not have bathrooms or kitchens" (Zheng et al. 2009, 435). Moreover, they report that urban villages, despite "their positive role in providing housing to migrants, . . . are of great concern to city governments. High crime rates, inadequate infrastructure and services, and poor living conditions are just some of the problems in urban villages that threaten public security and management" (Zheng et al. 2009, 426). In the villages surveyed by Zheng et al., migrants dwell in a "mean per capita living space" of 8.2 square-meters (2009, 434). However, "housing expenditure accounts for only 19.5 percent of migrant's household income," leading Zheng et al. to the conclusion that "from the housing affordability perspective these migrants can not only afford their current rental units . . . but actually can afford even better housing. Nonetheless, migrant workers do not choose to improve their housing consumption. Instead, they save almost all of their earnings, which are customarily sent back as remittances to their rural homes" (Zheng et. al 2009, 441).

This leads us to C. Cindy Fan and Wenfei Winnie Wang's insight that a more expansive conception of the "international circulation of labor and international labor markets" and a more expansive approach to overall migratory patterns are required if the subtleties of contemporary processes of reproduction are to be comprehended (2008, 207). Rather than simply emphasizing "net personal" economic gains to interpret migratory options vis-à-vis employment opportunities, a more multilevel understanding of reproduction, especially as it is manifested in rural-urban migratory patterns, is required. The contention is that reproductive strategies are not undertaken solely on an individual basis, but the unit that is the focus of reproductive decisions is typically the family instead of the individual. Such strategies are not solely undertaken with economic goals in mind: frequently, the primary focus is on the survival of family structures and *not* on the engrossment of individual savings. Of course, family structures and economic well-being are (or can be) intimately connected, but their connection can also be de-linked, at least to a certain degree.

While Fan and Wang are making this case about China, a country often thought of as consisting of individuals more influenced by communal concerns than the individual goals that provide incentives for those living in the West, it would be interesting to revisit, say, Italian immigration patterns to the United States in the late 1800s and early 1900s, to ascertain to what degree those patterns were also impacted by motivations that extend beyond purely individual concerns. This could go some way to debunking the myth of the individual utterly disconnected to anything else except his or her own immediate self-interest as the model par excellence of demography and economics. And though it may be true that, as Foucault says, "Migration is an investment; the migrant is an investor," an expansive definition of "investment" is needed to encompass the wide variety of motivations underlying modern migratory patterns, one that includes economic motivations as well as familial, political, and cultural motivations (2004, 230).

Fan and Wang, citing interviews conducted in 1995 and 2005 by China's Research Center for Rural Economy of the Minister of Agriculture, reference the following, which bolsters the contention that the reproductive necessities of the extended family play a major role in decisions regarding processes of reproduction: "Three years after Chen Guowei . . . began migrant work, in 1990, he was injured and became paralyzed. His wife's farming activity then became the only source of livelihood. She, at the same time, had to support the two sons' education, pay off a debt, and take care of the disabled husband. According to the couple, the family did not have enough to eat, until the two sons were old enough to do migrant work, the oldest son beginning at the age of 14 and the younger son at the age of 17" (Fan and Wang 2008, 234). So in this case, reproduction on an individual basis is subsumed within the exigencies of reproduction on a familial basis. The contingencies of a disability, the reversal of traditional gender roles, and the introduction of young people (children) into the labor market also play roles within a reproductive process more complex than the model offered by a sole individual economic unit (a person) striving toward personal enhancement vis-à-vis financial autonomy (self-interest).

To continue with the account of Chen Guowei's family: "This household intergenerational division of labor is mutually beneficial: migrants send back remittances that benefit the entire household, and grandparents that stay behind take care of the grandchildren so that the migrant workers can earn as much as possible to build up their economic security" (Fan and Wang 2008, 234). Part of the effort to earn as much as possible in order "to build up their economic security" consists of living as cheaply as possible while working in the city. Thus, the choice of renting living space in a crowded house in an urban village fits into an overall plan, which it is hoped will lead to the fulfillment of the dream of the parents to construct homes in the village for their children and their grandchildren. From this one can gather that an intergenerational division of labor may lead to *intergenerational* rather than *individual* reproduction. For such an intergenerational "dream" to come to fruition, all individuals must play their part. However, as already stated, the entire effort is postulated on a familial rather than an individual basis.

Referring to You-tien Hsing's *The Great Urban Transformation: Politics of Land and Property in China* can help to set urban villages and their role in China's housing nexus within some sort of context. Land speculation was instituted in China in 1988, when "the country's land leasehold market was formally established, thereby separating land ownership from land-use rights" (Hsing 2010, 5). A chronic shortage of housing in urban centers was coterminous with both a massive influx of rural migrants into the city and a profligate building spree, set loose by the combination of China's rapidly expanding economy, speculative desire (i.e., greed), and the perceived requirement to catalyze urban spaces into the sort of financial legibility (e.g., the skyscraper, the corporate headquarters) that is viewed as conforming with the status of world cities such as New York, Tokyo, and Paris. Urban villagers, making a transition from a livelihood based on agriculture to one based on the fact that they controlled land use within a certain segment of the enveloping city, leveraged their position within the metropolis.

If one focuses both on Fan and Wang's concentration on familial rather than individual interests as incentives for migration and Hsing's construal of urban villagers as participants in the wild speculations of China's burgeoning real estate market, one finds essential contradictions to the foundational principles of both capitalism and communism. On the one hand, self-interest, that Smithian cornerstone of classical economics, is trumped by the larger interest of the family, as migrants, organizing their reproductive activities along familial rather than individual lines, rent space in urban villages. On the other hand, villagers, having derived rights to lease out space from their status within a communal property model, lease out space in a thoroughly capitalistic mode of self-interest. So in the urban village, the two primary models of reproduction, the individual mode and the communal mode, come together and switch places. Still intact though nearly fragmenting, these reproductive modes subsist immediately adjacent to one another in China's inchoate communistic capitalism (or capitalistic communism, as the case may be).

And all this must be placed within the larger context of international economic competition, in which China is playing a larger and larger role, as well as within the context of intracompetition between Chinese cities to become one of the country's leading world cities. "Since the mid 1990s," writes George C. S. Lin, "China's urban spaces have been reproduced through a process of city-based and land-centered urbanisation in which large cities have managed to reassert their leading positions in an increasingly competitive, globalising and urbanising economy" (2007, 1846). China has also been economically reproducing itself at an unprecedented rate: "Its immense labor force" accounts "for some 29 percent of the world's total labor pool" (C. K. Lee 2007, 1). This labor pool has helped to catapult China into a leading position in the world's economy. "For 30 years, China has enjoyed average annual growth of about 10 per cent. In 1990, its income per capita was 30 per cent lower than the average for sub-Saharan Africa—today, it is three times greater, more than $4,000" (Zoellick 2011, 9). Such expansive growth is continuing, despite recent downturns in China's economy.

That this focus on urban reproduction and urban expansion has come at the cost of the loss of agricultural land is not lost on Lin, who reports that even before the most recent wave of urban expansion in the China of the twenty-first century, the greatest wave of urban expansion in the history of the planet, Chinese agricultural acreage was vanishing at an astonishing clip. "Official Chinese statistics indicate that, between 1978 and 1995, the total cultivated land shrank substantially from 99.39 to 94.97 million hectares" (2007, 1828). As China moves from a rural to an urban economy, it will be interesting to track how they secure that most vital element of reproduction: food.

Retrieving the topic of the urban village per se, Hsing relates the manner in which the party secretary of one such anonymous village, whom she renames as Mr. Deng, shepherds the interests of his particular domain. "*I asked Deng what his goal was as party secretary of the village. . . . his answer was brief and clear: to increase the value of the village-shareholding company from 300 to 500 yuan per share, and to successfully see through the redevelopment of Shuping village*" (2010, 122; italics, Hsing). While it may seem odd and even bizarre to hear a party secretary of a village in a nominally communistic state hitch his village's economic star to a rise in share values, it should be added that "share values of the shareholding company are central to the livelihoods of the villagers, after 98 percent of the village's land was lost to urban expansion" (Hsing 2010, 122). Reproduction must proceed on some basis, and with only 2 percent of their land remaining following the incessant onslaught of an omnivorous urbanism, Deng's strategy makes eminent sense.

Here I should dispel any notion that all Chinese urban villagers are sitting pretty on top of ever-increasing share values. Many urban villages have simply been bulldozed and their former villagers relocated as "the urban government uses its administrative and planning power to grab premium land from the village and forces villagers to move" (Hsing 2010, 191). In this process of expropriation and relocation, "villagers who live along the inner ring of the urban fringe are moved to the outer ring of the urban fringe where land is even cheaper; and villagers at the outer ring are relocated to even more remote areas at the rural fringe of the metropolis" (Hsing 2010, 191). These demolition and relocation "drives" are frequently met with resistance. "Between 1990 and 2002, an estimated 50–66 million peasants lost all or part of their farmland and homes to local government land grabs and development projects. Protests by aggrieved peasants have been on the rise since the late 1990s. In 2005, the Ministry of Land and Resources recorded 87,000 protests related to land grabs, a 6 percent increase from 2004" (Hsing 2010, 17).

Referencing David Harvey's *The Urbanization of Capital* (1985), Hsing goes on to note that "from the perspective of the expansionist urban government and development powers, the multiple rings of urban expansion create a chain of what urban geographers call a 'spatial fix'" (2010, 192). The positive effects of this chain for developers and real estate speculators (who often are city officials as well) creates a reciprocal negative chain of effects for the less fortunate. As the displaced villagers move to the urban fringe, they drive up housing prices and push other residents further out into the fringe, and so on and so forth. Here the deterioration of the average villagers' chances of adequately reproducing themselves align with the average capitalists' chances of reproducing themselves at a grander scale. The spatial fix works to the advantage of the capitalists fixing their capital in space while the same fix works to the disadvantage of the villagers losing their share of space.

Some remarks should be inserted here about the population density of these villages. In 1993, the government attempted to limit the upward expansion of houses in the urban villages to a "maximum of three stories or 80 m in height" (Hsing 2010, 128). Villagers, taking these regulatory regimes as cues to expand quickly before the new codes could be enforced, jumped into action and increased the height of their "houses" to eight and even ten stories. To grasp the scale of the density of these villages, note that Shuping Village, for instance, has "reached a population density of 174,450 per square kilometer," which is "about 200 times higher than the national average for urban areas." This can be compared to that of "New York County, the most densely populated in the United States" which stands at "27,267 residents per square kilometer as of 2007, about one sixth of Shuping's density" (Hsing 2010, 129).

Urban expansion in China has created the by-product of an exponential expansion of the reserve army of labor. "Between 1980 and 2003, somewhere between 50 and 66 million Chinese peasants lost all or part of their farmland and houses" as a result of urban extension into the countryside (Hsing 2010, 182). China has a "'floating population'" that "stood at 221 million in 2010 and is projected to increase to 350 million by 2050" (Fan 2011, 1). This "floating" surplus army of workers has settled into a pattern of more or less permanent circulation between the country and the city. Such an enormous amount of people becomes prime fodder for exploitation as they funnel into the immense ranks of low-wage workers that have attracted so many Western firms to China.

And such a process of reproduction through demographic patterns of more or less permanent circulation obviously destabilizes the Marxian dichotomy between the town and the country as well as the demographic assumption of a teleological account of migration as leading to some final destination. Reproductive patterns such as these can be found in many other places as well, especially in what is now called the South. "Research on rural-urban migration in Africa, Asia, and Latin America has indeed noted the prevalence of temporary, circular migration between home and the place of migrant work" (Fan 2011, 3–4).

A version of such circulation, a form of bilocality, is reflected in the situation of many Mexican immigrants to the United States. "Mexican immigrants to the U.S. . . . may continue to use the farmland in rural Mexico as an economic asset and a basis of household activities" (Fan, Sun, and Zheng 2011, 2). As C. Cindy Fan notes in "Settlement Intention and Split Households: Findings from a Survey of Migrants in Beijing's Urban Villages," such patterns seem to ensure the highest chance for successful reproduction for the members of this floating population. So here is migration as "not just a one-way move from the origin to the destination but as an activity that engages both origin and destination societies" (Fan 2011, 1). And here is also the basis for a housing market in China's urban villages resulting in the situation in which urban villagers are being "outnumbered" as much as 5:2 by rural villagers renting space from them. Such a form of reproduction accrues benefits to both parties, and to risk positing a tautology, as long as it does so it will continue to be a factor in Chinese reproduction.

After this brief survey of one aspect of the reproduction of the surplus army, I turn next to the middle class and the story of the rise and fall of their reproductive status in the United States of America.

Middle Classes

"Our best-grounded expectations of an increase in the happiness of the mass of human society are founded in the prospect of an increase in the relative proportions of the middle parts," writes Thomas Malthus in *On Population* (1960, 584–85). Approximately 125 years after the first edition of Malthus's *Essay on Population* was published in 1798, Henry Ford came to the conclusion that paying wages sufficient to turn his workers into consumers would not only create hundreds of thousands of new purchasers of the automobiles rolling off his assembly line, but would also have the felicitous by-product of quelling the militant passions of the American working class. Co-option of the worker through at least a *seeming* entry into the middle class—the purchase of a Model-T being the ticket to that status—was correctly identified by Ford as being both a boon to business as well as a bounty to the workers. The transformation of the working class into the middle class in the United States from 1920 to 1970 was one of the greatest economic tour de forces in the history of capitalism and seemed to deflate any need for a revolutionary movement in the United States along Marxist lines.

However, in 1973 the bottom dropped out of the American economy due to the realignment of various factors, including but not limited to the creation of OPEC (Organization of Petroleum Exporting Countries) and the Nixon Administration's move away from the economic ground rules drawn up at Bretton Woods in 1944. The decades following this transformation have witnessed a slow erosion of the middle class in the United States. "From the start of the long downturn [in 1973], real wage growth throughout the advanced capitalist world had progressively decelerated, and shrank toward disappearance as the 1990s progressed," writes Brenner in *The Economics of Global Turbulence* (2006, 193).

Brenner does not stand alone in such an analysis, nor are such comments confined to the retrospective viewpoint afforded by the passage of time. In *The Deindustrialization of America: Plant Closings, Community Abandonment, and the Dismantling of Basic Industry*, first published in 1982, Barry Bluestone and Bennett Harrison are already sounding the tocsin for the implosion of America's middle class: "By the beginning of the 1980s . . . the system that seemed so capable of providing a steadily growing standard of living during the turbulent 1960s had become totally incapable of providing people with a simple home mortgage, a stable job, or a secure pension" (1982, 4). In *The Great U-Turn: Corporate Restructuring and the Polarizing of America*, published in 1988, Harrison and Bluestone write, "In key sectors throughout the economy, workers have lost jobs numbering in the millions, and those fortunate enough to hold on to theirs have often had to submit to substantial reductions of wages and benefits" (1988, 35–36). In Paul Krugman's *The Age of Diminished Expectations: U.S. Economic Policy in the 1990s*, first published in 1990, Krugman reports that "the real wages of blue-collar workers have declined fairly steadily for the past decade" (1995, 27). Moving forward and focusing on Latino income, in *Magical Urbanism: Latinos Invent the U.S. City*, first published in 2000, Mike Davis reports on statistics gleaned from a January 2000 Federal Reserve study reflecting that "median household net worth among Latinos fell 24 percent" in the five-year period from 1995 to 2000 (2001, 121). Finally, in a *New York Times* editorial of October 7, 2011, the

Times leads off its "More Bleak Job Numbers" opinion piece with this asser-
tion: "It would take a lot of optimism to put a positive spin on the jobs report
for September, released on Friday by the Labor Department. . . . There are
still 14 million people officially unemployed, and nearly 12 million more who
have given up actively looking for work or who are working part time but
need full-time jobs" (2011, 1). Though unemployment in the United States
has fallen as of June of 2015 to 5.3 percent, this improvement certainly hasn't
made much of a dent in the long trend I have been underlining. The Long
Downturn has matched the Long Upturn in its effects on the overall economic conditions in the United States and the stability of its middle class, effects that have been at least somewhat concealed by bubbles of various kinds,
whether they be of the military spending, dot.com, housing, or credit card
debt variety.

  For instance, the real estate bubble of the early 2000s managed to veil the
overall downward trend of the post–Golden Era downturn, at least to a certain degree, but the Great Recession of 2008 ripped that veil asunder, open-
ing a raw and transparent window upon the inconvertible fact that for the
vast majority of Americans, and certainly for the great vaunted middle class
of the United States, the economic promise of the American Dream had dis-
integrated. Describing the housing bubble, Brenner states that "between 1997
and 2005, housing prices increased by 51 percentage points more than did
rental prices, an entirely unprecedented divergence. This is a clear indication
that housing prices are not being driven up by fundamentals—such as rising
incomes, population growth, or a change in consumer preference in favor of
housing—which would equally affect rental and housing prices, but by what
is in essence speculation" (2006, 320). Meanwhile, "aggregate real compensation in the private economy (real compensation per employee, multiplied by
employment) . . . increased between 2000 and 2005 at the lowest rate of the
postwar epoch—at half the rate recorded between 1980 and 2000" (Brenner
2006, 332). And the buying power of the middle class, despite positive gains
in the unemployment levels since the crash of 2008, manifests a continuing
trend to the negative, thinning out the ranks of the already thin American
middle class.

  What I want to do in this section is focus on the story of the rise and
fall of the American middle class. I recruit this as my present subject matter
for three reasons. First, it complicates Marxian class analysis by torquing the
definitional boundaries of the working class and the middle class, for as the
working class in the United States gained entry to at least an ostensible status

as bona fide members of the middle class during the boom years of the 1950s and 1960s, the integrity of Marx's strict demarcations between the middle and working classes seemed to fray, straining Marxian class analysis while simultaneously bolstering a kind of utopian vision of capitalism in which all boats rise from here on in to eternity. Secondly, I use the decline of the American middle class as my focus because its endgame seems, at least from the perspective of the present moment, to swivel upon a globalized redefinition of the middle class, one that pivots away from the constraints of "our own Anglo-American society," as it incorporates a newly constructed middle class of consumers from the BRICS (Brazil, Russia, India, China, and South Africa). Such an incorporation has the tangential, serendipitous by-product of serving our long-term project of globalizing this present study. Finally, the story of the ascension and declension of the American middle class allows me to examine a case in which the requirements necessary for reproduction as a member of a certain class become so strained that they snap, leaving those whose reproductive status as members of a certain class had seemed to be secure dangling in a bankrupt void.

Social ascension ("climbing") in the United States through the acquisition of symbols of status did not originate in the postwar boom of the 1950s.

> In the Connecticut River Valley, threatened elites of the 1760's and 1770's enacted "a drama of class dominance" by either building Georgian homes in the style of public buildings, thus symbolically conferring public authority on their private spaces, or, less expansively, constructing elaborate scrolled and pilastered doorway facades over rather ordinary homes as barriers against threats to the "natural" social order. However, the artisans employed to create these class markers themselves acquired so much liquid capital that by "1768 a tailor, a shoemaker, and a joiner has placed elaborate doorways on their own houses." (Fliegelman 1993, 112)[4]

What is reflected in this example can serve as a foreshadowing of a process repeated throughout the history of the United States (and throughout the history of capitalism itself, for that matter), as status symbols are developed by the upper class, maintained as markers of social and economic prestige for a more or less extended period of time, and then "purloined" by other classes as symbols of their own ascending status. "The lower groups often imitate the trends, styles, fashions, and practices set by their betters; the stratification system maintains and nurtures itself by this imitation" (Fernandez 2003, 100). Typically, however, by the time such markers of Bourdiean cultural and social capital have trickled down to the middle and lower classes, new markers have been formulated, produced, marketed, acquired, and set into position as signs of status by the upper class, leading to a continuous looping action of status symbol acquisition and appropriation, which bolsters primary processes underlying both capitalism and class demarcation, drivers of a consumeristic perpetual motion machine.

Jumping from the Connecticut River Valley of the 1760s to the American suburban housing development of the 1950s may seem like a vault too far to be performed; however, what is discovered in the explosion of home buying during the 1950s and its "necessary" corollary, an explosion in the purchasing of household durables, serves as a distant mirror to the same processes that were occurring in the Connecticut River Valley in the 1770s: that is, liquid capital being exchanged for symbols of status.

The story of the 1950s suburban home can be told via a manifold of narrative devices and from perspectives as dissimilar as feminist wrath at the cloistering of women in kitchens and supermarkets to the hailing of those same women for staying in the kitchen and shopping at the supermarket by conservatives nostalgic for the "golden age" of the (white) middle-class American family. However, I want to stick with a more constrained economic story, so let me begin with President Herbert Hoover's response to the first wave of the Great Depression.

Though Hoover "called for accelerated construction spending by public authorities and for increased capital outlays by the major private investors," in order to boost the economy, neither the sufficient amount of liquid capital nor the public institutional financial infrastructure to administrate such an outlay were in place to deliver any relief in the form of say, low-interest loans for home-buyers" (Barber 1985, 189). Housing construction declined throughout Hoover's term as president, despite perorations from the White House that were intended to stimulate home construction, such as Hoover's remarks in March of 1930 that "one direction which is always economically and socially sound is in home building, in which there is large consumption of labor directly and indirectly through producers' and consumers' goods. Increasing improvement in housing conditions is of the utmost social importance" (Hoover, qtd. in Barber 1985, 96),[5] and the creation of the Home Loan Discount Bank in July 1932. Hoover had envisioned a market-oriented solution to the Depression, with every American family purchasing a home and every home outfitted with durables such as the already tried and true automobile as well as new-fangled items such as washing machines, vacuum cleaners, refrigerators, and so on. Hoover had the right idea (that is, if rightness coincides with capitalism practiced on a mass consumer scale), but the wrong time. Twenty years later, his vision was percolating along quite nicely, as America's economic predominance following World War II and the creation of such institutions as the Federal Housing Authority (in 1934) and Fannie Mae (in 1938) created perfect financial vehicles for individual (familial) economic units to use in order to acquire prime economic drivers such as houses, cars (two in every garage), stoves (first gas-fueled and then electric ranges), televisions, swimming pools, and so on.

The establishment of the ambience in which such levels of mass consumption could flourish led to the "golden years" of American suburbia in which the working class of the United States was transformed into its middle class. Status symbols such as private homes, automobiles, and color TVs were no longer items restricted to the upper class. Just as the shoemakers, joiners, and tailors of the Connecticut River Valley of the 1760s and 1770s performed their entrance into the middle class through the artisanal elaboration of their doorways, factory workers, plumbers, and schoolteachers of the 1950s and 1960s were able to perform at least the appearance of being solid members of the middle class by elaborating their economic position through the purchase of mass-produced markers of middle-class status. That these markers could never catch up with the symbolic capital of the wealthy is beside the point (and, indeed, as previously noted, this is a primary function of the entire reproductive process of both capitalism and the class system that capitalism both feeds off of and fosters). The point is that—at least for a few decades—the working class counted itself and was tallied by others as being certified members of the middle class.

If this indeed is not class structure as formulated by Marx and Engels, it also contrasts sharply with other classical formulations of the middle class, such as Tocqueville's and Weber's. I quickly survey those respective formulations and the types of workers upon which they were based, simply to demonstrate how far removed Tocqueville's and Weber's formulations are from the constitution of the middle class of America during the mid-twentieth century.

In discussing the pre-revolutionary era in *The Old Regime and the French Revolution*, Tocqueville states: "Most official posts were staffed by men belonging to the middle class, which had its own traditions, its code of honor, and its proper pride. This was, in fact, the aristocracy of the new and thriving social order which had already taken form and was only waiting for the Revolution to come into its own" (1955, 63). Here at one and the same time is a statement that could be construed to prefigure Weber's insertion of the middle class into a slot labeled "bureaucracy" ("official posts") as well as a confirmation of the widely accepted idea that the French Revolution was an upheaval that vaulted the bourgeoisie into power. Further on in *The Old Regime*, Tocqueville assigns a tighter definition to the pre-revolutionary middle classes, identifying them with the legal system: "One might almost say that the whole middle class was concerned in one way or another with the administration of justice" (1955, 193).

This, of course, does not square with the American definition of the middle class circa 1960, when carpenters, plumbers, and stevedores were tallied as bona fide members of the middle class. However, this may also testify to a more strictly egalitarian and more purely economic (and capitalistic) American definition of the middle class, in which a sole reliance on quantity of income is used as criteria for entry into the middle class rather than such items as family lineage or profession. But then neither does it square with Tocqueville's own description of class structure in *Democracy in America*. In the United States, at least as witnessed by Tocqueville in the 1830s, a middle class practicing the crafts of government and administration was conspicuous by its absence. "Nothing strikes a European traveler in the United States more than the absence of what we would call government or administration. . . . The hand directing the social machine constantly slips from notice" (1969, 72). But then in Tocqueville's version of America the rich and the poor are pretty much absent as well: "I realize that among a great people there will always be some very poor and some very rich citizens. . . . As there is no longer a race of poor men [in America], so there is not a race of rich men; the rich daily rise out of the crowd and constantly return thither" (1969, 635).

Here it is a bit difficult to decipher whether Tocqueville was merely too blinded by the "democratic" expanse of the American scene so as not to be able to perceive the class stratifications already in place in the mid-1830s or whether the United States has changed so dramatically since that "egalitarian" era that his depiction of an America without either the rich or the poor strikes one as so alien as to be fatally flawed. Perhaps the truth lies somewhere in between those two poles. Or perhaps what appears to be a misperception of the American scene vis-à-vis class structure is, on the part of Tocqueville, simply a clear statement of his belief that, as Poggi articulates the point in his study of the Frenchman, though "economic disparities between individual citizens may be great [within the United States], . . . the population at large shares a relatively high degree of economic security and a relatively comfortable standard of living. It thus shares a feeling of partaking, however diversely, in a vastly successful cooperative endeavor, an increasing mastery of the whole people over its environment" (Poggi 1972, 56). However, even retrospectively it is difficult to accept this Panglossian estimation of the United States as a "cooperative endeavor" sustaining a "relatively high degree of economic security." To accept this as an accurate portrayal of the United States today would be ludicrous.

Weber's contribution to the theorization of the middle class was not simply his identification of the middle class with the members of the bureaucracy, but also his insertion of the importance of *social* esteem and *status* markers as equal in significance to *financial* status and *economic* markers in the formation and maintenance of that class. "Whether he is in a private office or a public bureau, the modern official . . . always strives for and usually attains a distinctly elevated *social esteem* vis-à-vis the governed" (Weber 1968, 959; italics, Weber). With the modern official's "social position . . . protected by prescriptions about rank order and, for the political official, by special prohibitions of the criminal code against 'insults to the office' and 'contempt' of state and church authorities," the status of middle-class professionals and officeholders was also buttressed by the creation of a "system of specialized examinations or tests of expertise (*Fachprufungswesen*) increasingly indispensible for modern bureaucracies" (Weber 1968, 959, 999). And since "office management . . . usually presupposes thorough and expert training," such training becomes a prerequisite of attaining positions within the hierarchy of office management of whatever kind (Weber 1959, 198).

Tracing bureaucratic structuration to "the consistent patrimonial-bureaucratic administration" (Weber 1959, 1044) of ancient Egypt and bureaucrats themselves to Egyptian scribes, Weber links the "middle class" functions of knowledge storage and decree regulation with a certain amount of power. In such a conception, both entrance into as well as the function of the middle class appear to be completely at odds with the condition of the American middle class. However, much of Weber's analysis can be retained once it is recalled that union membership and certificates of expertise (for schoolteachers to court reporters to auto mechanics) are frequently an entrance requirement into the ranks of the American middle class, and that despite the flood of information made available by the Internet, bureaucrats still hold key positions within the systems that store and disseminate knowledge.

Marx and Engels took a very different view of the middle class. In *The Communist Manifesto*, Marx and Engels identify "the lower strata of the middle class" as "the small trade people, shopkeepers, and retired tradesmen generally, the handicraftsmen and peasants," but then they make the claim that all these are "sinking gradually into the proletariat, partly because their diminutive capital does not suffice for the scale on which Modern Industry is carried on, and is swamped in the competition with the large capitalists, partly because their specialized skill is rendered worthless by new methods of production" (Marx 1994, 165).

However, this is not what occurred in the United States in the 1950s, as members of the proletariat rose into the middle class. That this may have been only a temporary ascension, confined to one "golden era" in the history of capitalism, does not make it any less interesting from an analytical outlook. What happened to turn at least one feature of Marxian analysis on its head? I return to this question after a brief survey of the definitional criteria of the middle class, according to Marx and Engels.

Marx and Engels seem to split the middle class in two. On the one hand, there are those who are proximate to the upper class, referred to in *The German Ideology* as the "big bourgeoisie . . . big merchants and manufacturers," and on the other hand, there are the "petty bourgeoisie" that had been "concentrated in the guilds" during the feudal mode of production (Marx 1994, 138). The petty bourgeoisie, according to *The Manifesto*, "fight against the [big] bourgeoisie, to save from extinction their existence as fractions of the middle class. They are therefore not revolutionary, but conservative. Nay more, they are reactionary, for they try to roll back the wheel of history" (Marx 1994, 167). It's difficult to discern what the "middle" of the middle class consists of in a Marxian analysis. Those who fall in between the "big" and the "petty" bourgeoisie seem to be either in ascension or declension, slipping through the cracks and falling into the ranks of the working class or groping upwards into the status of the "big bourgeoisie," with their sights set on perhaps eventually gaining entry into the ruling class.

Now that the ways in which Tocqueville, Weber, and Marx and Engels conceive of the middle class have been sketched out, howsoever slightly, the startling nature of the American middle class circa 1960 can begin to be appreciated. These were essentially proletariats, workers who had risen into the middle class. They certainly were not members of the judicial, administrative, or bureaucratic orders of society. In no way should the impression be left that this massive ascension was simply due to largesse on the part of the owners of industry. Ford's initial insight that mass producers must also be mass consumers did help in creating the conditions for a takeoff of the

American middle class, but the demands and struggles of workers, especially during the 1930s, also played a crucial role in its ascension. Labor peace was guaranteed and profits were secured, as the government managed to catalyze an economic "miracle" through assistance programs such as GI college loans and federally guaranteed mortgages.

Skipping over any further discussion of the Golden Era's collapse while also cutting short any further narrative detailing the succeeding forty-some years of middle-class economic disintegration, yet still keeping in mind those two events as prime milestones in the larger story here, I want to focus on a series of articles that appeared in the news media in the summer of 2011, articles that forecast a new international constituency of the middle class while also articulating the dispensability of the American middle class in the economic planning of corporations.

On August 8, 2011, Don Lee of the *Los Angeles Times* reported, "Many major U.S. companies are making big plans to expand overseas even as some of them announce new layoffs at home" (2011, 1). Being that such "big plans" had been pretty much typical since the dawn of American deindustrialization in the early 1980s, why was this newsworthy? Because this was consumer oriented rather than production oriented, with markets rather than factories being located overseas. Lee explains why U.S. companies are making such moves now: "They're beginning to give up on the American consumer as a source of future growth. . . . In effect, as many corporate executives look ahead, the United States has a diminishing place in their thinking" (2011, 1). Lee cites as an instance of this trend "one of the biggest marketers of children's car seats," Newell Rubbermaid, Inc., which is "expanding in Brazil instead of the United States" (2011, 1). And the stated reason for this is that "while young Americans are putting off having children, in part because of the poor economy, Brazil's middle class is growing, and many more young couples are starting families. So more Brazilians have the money to buy new, upscale car seats while more U.S. parents are making do with cheaper brands or hand-me-downs" (D. Lee 2011, 1).

This growth in South America's middle class is also reflected in a *Los Angeles Times* article of March 9, 2012, "Latin American Air Travel Soars," in which Chris Kraul, reporting from Bogota, informs us that "the region [Latin America] led the globe in air travel last year, with 10.2% more passengers than in 2010, more than double the 4% jump in North America," according to statistics gathered by Kraul from the International Air Transport Association (2012, B1). "Stoking the expansion is a boom in Latin American exports, including oil, coffee, copper, and soybeans. That has boosted incomes and helped expand the middle class," which, in turn, has served as a fillip for air travel (Kraul 2012, B1). This booming growth rate bears directly on middle-class reproduction, on both the scales of the individual and the class, bridging the gap between the connotations of the word; generative *sexual reproduction* in the United States, Lee suggests, is not occurring at least at previous rates, due to a sluggish economy, and, therefore, *class reproduction* is effected.

So here it might be stipulated that while the first post-Fordist wave of late capitalism instantiated production on a global scale, a second wave is instantiating consumerism on a global scale.

In the targeting of potential customers, companies such as Newell Rubbermaid are not simply abandoning the U.S. market, as the domestic market remains "stable" and "remains huge, but it's not growing significantly and prospects—reflected in the downgrading of the nation's debt . . . —are similar for the next few years" (D. Lee 2011, 1). However, markets such as Brazil, India, and China are expanding exponentially and are, at least to a certain extent, replicating the consumeristic process that the United States went through in the 1950s and 1960s. Millions upon millions of newly instantiated members of the middle class want durables as markers of their recently attained status: refrigerators, dishwashers, one car for mom and one for dad, car seats for the children, and so forth.

U.S.-based firms are simply responding to economic signals: when economies such as Brazil's are setting a pace to expand two to two and a half times faster than those in the United States and other industrial nations, of course companies will zero in on them as potential customers. And the economic turmoil roiling the United States and the European Union could lead to a long-term refocusing of the market, in which the (former) industrialized nations are caught in a financial tailspin powered by perpetual negative feedback loops. "A downturn in the rich countries [the United States and the members of the EU], of course, would dampen growth around the world and could lead to a new global recession. Gripped by such worries," companies could be motivated "to pull back further in the U.S. and look even harder at investing overseas to protect their profit margins" (D. Lee 2011, 1).

Lee's assessment is buttressed by an article written by his colleague, Henry Chu, and published in the *Los Angeles Times* on August 21, 2011: "Emerging Economies Faring Better in Downturn." "Look at the state of the economy from anywhere in America or Europe these days and all is gloomy: Governments deep in debt. Consumers reluctant to spend. Businesses afraid to hire. But gaze out from the vantage of some of the world's emerging economies and the picture gets brighter" (Chu 2011, 1). And even though China's economy has slowed down from its 10 percent annual rate of growth to a still strong 8 percent growth rate, that far exceeds Germany, the "shining star" of the developed countries, whose "economy grew by just 0.1% in the second quarter" of fiscal 2011 (Chu 2011, 1). In fact, in "2009 the PRC [People's Republic of China] overtook Germany to become the world's largest exporter of goods, with 34 firms in the Fortune 500," with "the market capitalization of Chinese firms in the FT [*Financial Times*] 500 . . . second only to that of American firms, while in the banking sector, the top three positions were occupied by Chinese institutions" (Nolan and Zhang 2010, 97). However, Nolan and Zhang qualify this when they state, "The international operations of China's leading banks remain far behind those of the Atlantic core. China does not have a single bank among the world's top fifty, ranked by geographical spread" (2010, 106, 107). So China still has some distance to cover before it is a bona fide member of the core, at least according to an analysis based on Nolan and Zhang's criteria.

Returning to Brazil, its burgeoning relationship with China is helping to create what is beginning to look like a Latin American economic juggernaut: "Brazil is now an agricultural superpower and natural-resources giant, shipping vast quantities of commodities to China, the world's workshop" (Chu 2011, 1). But Brazil isn't just supplying China and other countries with raw materials and agricultural products. On August 22, 2011, United Press International announced that Brazilian aircraft manufacturer Embraer had "delivered the first of its E-190 jets to China's CDB Leasing Co., Ltd. . . . CDB Leasing Co., also known as CLC, has ordered 30 Embraer E-190s, including 20 firm orders and 10 options, all of which will be operated by China Southern Airlines" (UPI 2011, 1). And while twenty "firm" orders with ten options as an add-on does not amount to a major manufacturing agreement, it does signal the potential for such an agreement, the possibility of which further corrodes the West's hold on high-end products and the middle-class incomes that go along with manufacturing such products.

Despite the numerous misfires of Marx and Engels's prognosticative powers, precisely such a turn was predicted in *The Communist Manifesto*. As Michael Watts points out in his Hettner-Lectures of 1999: "Whatever its predictive failures as regards politics and revolution, the Manifesto was remarkably prescient as regards capitalism expansion, circuits of accumulation and what we would now call globalization" (2000, 52). Watts quotes the *Manifesto* to buttress this claim: "The need for constantly expanding markets chases the bourgeoisie all over the surface of the globe" (2000, 52).[6] The fact that a communist revolution first happened in the backwater of Russia instead of the industrial powerhouse of Germany would have surprised Marx and Engels, but the fact that capitalism has spread its tentacles over nearly every corner of the globe would not have shocked them in the least. The concomitant fact that different national subsets of humanity are replacing others as members of the middle class would not have surprised them either, as capitalism requires "constantly expanding markets" and the customers to feed those markets as well. Whether they be Americans, Chinese, Brazilians, Kenyans, Peruvians, or Outer Mongolians makes not a whit of difference to the essential requirements of capital expansion.

The demise of the American middle class and the ascension of the middle class in the BRICS (though Russia's middle class has lagged behind) has not gone completely unnoticed by the American punditry. In his Labor Day column of 2011, E. J. Dionne of the *Washington Post* wryly suggests that the holiday's name be changed to "Capital Day" in order to commemorate a society in which "we tax the fruits of labor more vigorously than we tax the gains from capital . . . and we hide workers away while lavishing attention on those who make their livings by moving money around" (2011, 1). Dionne's colleague at the *Post*, Harold Myerson, references a "stunning study written by Michael Greenstone and Adam Looney of the Hamilton Project, published in the Milken Institute Review, [which] reveals that the median earnings of men ages 25 to 64 declined 28 percent between 1969 and 2009" (2011, 1). In their study Greenstone and Looney add, "The median wage of the American male has declined by almost $13,000 after accounting for inflation in the four decades since 1969. (Using a different measure of inflation suggests a smaller, but still substantial, drop in earnings.) Indeed, earnings haven't been this low since Ike was president and Marshal Dillon was keeping the peace in Dodge City" (Greenstone and Looney 2011, 12).

I conclude this section with the case of one German Morales, a house painter who "emigrated from El Salvador [to Northern Virginia] at age 12 in the late 1980s" and by 2005 seemed to have earned the status markers of entrée into the middle class, with purchases of a "personal watercraft" and a home in Woodbridge, Virginia (Saslow 2011, 1). However, by 2007, Morales and his wife, Illiana, lost their house to foreclosure (a house financed by one of the notoriously toxic adjustable-rate mortgages that partially led to the collapse of the American economy in 2008) and drained their son's college fund of $35,000. Morales was reduced to staying busy by painting and repainting his new house until work picked up, which also served as a way to maintain his skill level as a nominal member of the reserve army of labor. "When he was on a job, he sometimes painted past midnight to compensate for his lack of manpower. When he had nothing, he drove from Richmond to Baltimore to drop off business cards, sometimes pulling over to nap in his truck" (Saslow 2011, 1). And so one might want to add that even the immigrant's version of the American Dream as well as the middle-class version of that dream is in danger of toppling.

The Ruling Class

In this section, I focus on the reproduction of the world's elite of this brave new economy, the top-one-percenters who have been exponentially increasing their financial returns during the last few decades while the middle class shrinks and the lower class expands. I especially want to highlight several projects, either newly built or in the planning or construction phases, occurring in China, Egypt, Saudi Arabia, and London. First, though, let me reference Marx and Engels on the ruling class. In *The Communist Manifesto*, the ruling class is figured as "the modern bourgeoisie," a "most revolutionary" class, "itself the product of a long course of development, a series of revolutions in the modes of production and exchange" (Marx 1994, 160, 161, 160). The advance of capitalism runs parallel to or, perhaps better, is linked inextricably with the advance of the bourgeoisie. In its relentless drive to reproduce wealth and reproduce itself as the holder of that wealth, the bourgeoisie "has resolved personal wealth into exchange value, and in place of the numberless indefeasible chartered freedoms, has set up that single, unconscionable freedom—Free Trade" (Marx 1994, 161). This expropriation of freedom and the substitution of "Free Trade" for "the numberless indefeasible chartered freedoms" leads to a situation in which "liberty" is understood as the freedom of the bourgeoisie to exploit the working class and to do so in the manner depicted by Marx and Engels in their allusion to Hobbes's famous invocation of the original state of nature: "naked, shameless, direct, brutal exploitation" (Marx 1994, 161).

A singular focus on accumulation is necessary for the capitalist class if they are to reproduce themselves. "The essential marxian [*sic*] insight . . . is that profit arises out of the domination of labour by capital and that the capitalists as a class must, *if they are to reproduce themselves*, continuously expand the basis for profit" (Harvey 2002, 116; italics, mine). After making this remark, Harvey follows up with a slight feint, stating, "We thus arrive at a conception of a society founded on the principle of 'accumulation for accumulation's sake, production for production's sake'" before realigning the argument by connecting it to the reproductive imperative: "This may sound rather 'economistic' as a framework for analysis, but we have to recall that accumulation is the means *whereby the capitalist class reproduces both itself and its domination over labour*" (Harvey 2002, 116; italics, mine). Reproduction is thereby retrieved back into the center of Harvey's analysis, right where it belongs, in my estimation.

Marx didn't intend his attack upon the bourgeoisie to be misunderstood as an attack on bourgeois individuals per se, for he was a materialist to such a degree that he viewed members of any class as subjects of historical forces well beyond their control. In the preface to the first German edition of *Capital*, Marx writes, "My standpoint, from which the evolution of the economic formation of society is viewed as a process of natural history, can less than any other make the individual responsible for relations whose creature he socially remains, however much he may subjectively raise himself above them" (1996, 10). Engels himself was the scion of a prosperous factory owner, which fact certainly didn't prevent him from being the cofounder of a movement quite counterproductive (*anti-reproductive*, that is) to his own class interests, to say the least. And, secondly, the inclusion in this statement of the categories of "capitalist" and "landlord" goes some way to differentiating the "big" bourgeoisie identified with the middle class in *The German Ideology* and the bourgeoisie as identified in *The Communist Manifesto*. In the former, the big bourgeoisie had been matched with "big merchants and manufacturers" (1994, 138). In *Capital* the capitalist and the landlord have taken their place, revealing perhaps a more nuanced analysis of who is actually constitutive of this class. Some of this may be due to the evolution of Marx's thinking, as *The German Ideology* was first published in 1845 while the preface to *Capital* was first published in 1867. Still, the individual member of the bourgeoisie could take little solace from Marx's caveat in the preface to *Capital* when in the very same preface he prophesizes thusly: "In England the progress of social disintegration is palpable. When it has reached a certain point, it must re-act on the Continent" (1996, 9).

But more simply put, in part 7 of volume 1 of *Capital* ("The Accumulation of Capital") Marx says that "we treat the capital producer as owner of the entire surplus value, or, better perhaps, as the representative of all the sharers with him in the booty" (1994, 565). This fragmentation of "the booty" to its "sharers" is derived from that capital producer "who produces surplus value—i.e., who extracts unpaid labour directly from the labourers, and fixes it in commodities, [who] is, indeed, the first appropriator, but by no means the ultimate owner, of this surplus value. He has to share it with capitalists, with landowners, &c., who fulfill other functions in the complex of social production" (Marx 1994, 564). Still this leaves out the landlord who is not involved in the splitting up of surplus value immediately attendant upon its extraction of profit by the capital producer. Those who profit from what Marx calls absolute rent or a monopoly upon a property (meaning that any discrete piece of land must have a particular owner or owners in the private property regime typically subtending a capitalistic mode of production) must also be inserted into the ruling class. Marx's theories of absolute rent have caused much confusion (a confusion I have neither the time nor the expertise to untangle); it is sufficient for my purposes to merely stipulate that people who profit from such property arrangements must be included in that class to which Marx confirms that he does not paint in a "*couleur de rose*" (1996, 10). Included in that class are also all those whom Shakespeare's Falstaff declaims have the "good wit" to "make use of anything" to turn a profit and will even "turn diseases into commodity" (*2 Henry VI*, 1.2.248), that is, those who know how to turn crises to advantage, such as those who sell umbrellas in a rainstorm at marked-up prices or those who force-feed austerity packages to debt-ridden countries. With that cursory and all-too-brief depiction of the ruling class in hand, let me turn to Dreamland.

In her description of Cairo's Dreamland, Farha Ghannam reports that this luxurious housing development, which includes its own "theme park [Dreampark], golf courses, tennis courts, horse-racing tracks, residential areas, conference facilities, a hospital, health resorts . . . a hotel . . . shopping centers," and its own television station (2008, 271), was constructed "to provide a mini-America on the Egyptian desert" (274). Though Ghannam warns her readers not to buy into a narrative that reduces this desire for a "mini-America" to a mere story of "notions such as 'neocolonialism,' 'McDonaldization,' 'cultural imperialism,' or 'Americanization,'" as "not all classes in Egypt desire these forms or aspire to acquire them," it is difficult to avoid reducing the Dreamland case into just such an instance of precisely such notions (2008, 274). This seems especially to be the case when we factor in Ghannam's own contention that "projects like Dreamland are based on what

I call the 'refusal to indigenize.' They reject the Egyptianization, Arabization, or Islamicization of the new forms they are introducing" (2008, 272). That the "refusal to indigenize" is moving forward by appropriating simulacra of America does not negate the integrity of this refusal; it simply demarcates it as one more instance of reproduction via replicating whatever the dominant cultural model happens to be at any particular point in time, whether it be the Romans imitating the Greeks circa 200 BCE, the Americans imitating the English in the 1700s, the Russians imitating the French in the 1800s, or the Egyptians and the Chinese imitating the Americans in the 2000s.

And it certainly does not demarcate the first instance of replication of metropolitan forms by a colony or a former colony, as this once constituted a primary component of the grand "civilizing" mission of colonialism and still constitutes a primary component of the grand modernizing mission of development.

However, it is important to clarify that such imitation of the West did not abruptly arise with the planning of Dreamland in Egypt. And disagreements about such imitative forms did not arise with the residents of Dreamland and their "refusal to indigenize." This is well-worn territory, playing out across many lands as Europe, the United States, the former USSR, and now the North have re-created their forms across the colonies, their respective spheres of influence, and now the South, with various constituencies battling over the meaning of "our." This has led to surprising and counterintuitive results, especially as the "creative class" in the North has recently come to valorize indigenous cultural forms.

Indeed, when Forrec, the American architectural design firm hired to create Dreamland, "suggested 'ancient Egyptian archeological themes, the Egyptian manager of the project 'preferred a North American Style . . . influenced by the kinds of Hollywood television programs and movies that most of the world watches and understands'" (Ghannam 2008, 273).[7]

Now, while it may be tempting for Westerners to denigrate as "pretentious" *any* Egyptian appropriation of a "North American Style," or to mock such an appropriation as yet one more example of a classical faux pas on the part of the parvenu, the right to even the tackiest of Hollywood fashions should not be sectored off as something that only "real" Westerners can "enjoy." It is also exactly what has been done on the part of many Westerners as they have attempted to ascend the hierarchy of status. Recall that those homes in the Connecticut River Valley circa 1776 were modeled after the latest in Georgian design, and that people in the "Mother Country" may well have looked askance at the pretentiousness of American colonists believing that they could assume such specifically English markers of class distinction.

The ubiquity of Hollywood products and the desire for marks of a Southern California lifestyle are not confined to high-end developments on the outskirts of Cairo such as Dreamland, as well as other Cairo developments such as California and Beverly Hills (that's right: California and Beverly Hills). Pico Iyer, wandering through the upscale suburb of the Zona Sur in La Paz, found a karaoke parlor called America and a shopping mall called San Diego. Indeed, "as I sat one night in a pizza joint, which boasted prices higher than Miami, a high-school girl at the next table, soignée as Catherine Zeta-Jones, shut her eyes and sang along, transported, to 'Hotel California' on the sound track" (2004, 96).

This transposition of the West onto the "rest" is occurring in the East as well. Lifescapes International, a landscape design firm located in Newport Beach, California, has created gardens for such places as the Mirage Hotel, the Bellagio Hotel, and the Wynn Resort, all in Las Vegas, as well as the Grove shopping mall in Los Angeles. The landscape design firm is now finding much of their opportunities in China. For instance, they are now designing the landscaping for the massive Greentown Orchid Project in Hangzhou. Commenting on this project, Lifescapes International's CEO Don Brinkerhoff says, "Thousands of years ago, the Eastern gardens first appeared in the West. We will try to implement classical European gardens back to the East" (Lifescapes 2011). So here we have the notion of cultural reciprocity deployed in a trope that vaults both time and space: just as "thousands of years ago" Eastern gardens "first appeared" in the West, so we in the West (in this case, Lifescapes International, Newport Beach, about fifty miles southwest of Hollywood) will now "implement" European gardens "back to the East." So a horticultural form that went through many transformations on its journey from ancient China to Europe to the West Coast of the United States is now being shipped back to China.

Though there may be a strong urge to dismiss or even mock this alleged history, the invocation of an almost karmically hued transcultural debt that is being repaid by Lifescapes International's landscaping of the Greentown Orchid Project perhaps should not be deflected so lightly. Is it not another instantiation of this universal desire of the global elite to frame their economic transactions in a cultural mode? And does it not also laminate what is essentially an urge to appropriate a North American Style (aka Hollywood) into that more distinctive marker (for Americans at least) of Europe? Just as the residents of the Connecticut River Valley looked to Europe for design inspiration, which would have the vital by-product of verifying their social capital, Lifescapes International is cloaking their landscaping aesthetic in a European veneer, which may serve to verify to the designers at Lifescapes as well as to their clients a signification of social capital extending from China to Europe to the Eastern littoral of the USA and eventually to Southern California and then back to China, albeit in a bastardized form. So simultaneously this serves as a symbol of China's desire for something American (and European) as well as America's desire for something European (and the eighteenth-century European desire for something Chinese).

So this form of design, evoking the American-Hollywood-Dreamland style, reproduces, or at least attempts to reproduce, status markers (homes, gardens, and real estate developments) that signify inclusion within a certain valorized milieu, a milieu perceived to be exemplary of the United States, which, despite whatever economic and political crises it may be undergoing, is still assumed by many around the globe to be the lodestar of status.

A very different method of conferring status is at work among London's supra-wealthy. In the August 31, 2011, edition of the *New York Times*, Sarah Lyall reports, "In a city that has some of the richest people and most expensive real estate in the world, well-off homeowners who have exhausted the traditional methods of home expansion—build up or build out—are enthusiastically branching out the only other way possible: down" (2011, 1). Chthonic household designers and architects are also breaking new ground, if you'll pardon the pun, in the fabrication of reality, that is, in virtuality: "One London homeowner," Lyall informs us, "installed an outdoor camera that projects real-time images of the changing sky onto the pool ceiling . . . around his new basement swimming pool . . . to compensate for the lack of natural light" (2011, 1). Yet another underground pool includes "a film of moving sharks . . . projected on the walls while the 'Jaws' theme song plays, and another covered in handmade golden ceramic tiles embedded with tiny lights that twinkle and give swimmers the feeling of being enveloped in the night sky" (Lyall 2011, 1).

So here, in London, in what could be called the metropole of metropoles, there seems to be a frantic desire to flourish in a style whose social capital exceeds minting, for here are status symbols so outlandish that their very creation, or, rather, *the very desire for their creation*, becomes a token of curiosity and wonder. That is, the pressing question becomes not "My, I never knew you had such a large amount of money, how did you get it?" but "How in

God's name did you ever conceive of spending your money in this peculiarly spectacular way?" And maybe that's the point: the spectacle of reproductive opulence supersedes itself so much that "simple" luxury becomes unworthy. One's wealth must be so ostentatiously displayed that money becomes not merely a nonfactor, but utterly worthless, the spending of it accomplished on such a gargantuan yet frivolous scale that its value depreciates to nothing. And what is this done in emulation of? Where is the model upon which this proceeds? Whose form and style of reproductive nonchalance does such excess descend from? Or is it merely objectionable because it's done in such bad taste? Here I must make a personal intervention to notate a member of the class of the blissfully eccentric rich with whom I was once acquainted who repainted the colors of her Alexander Calder mobile because the original hues did not match the color scheme of her Santa Barbara living room!

If one consults Marcel Mauss's description of the display of wastage that seems to be much the point of these displays, we may gain some insight into these "imaginative" feats of spending. Referencing "the Tlingit and Haida of Alaska, and the Tsimshian and Kwakiutl of British Columbia" (1967, 22), Mauss writes that during the winter, "on the occasion of a marriage, on various ritual occasions, and on social advancement, there is reckless consumption of everything which has been amassed with great industry from some of the richest coasts of the world during the course of the summer and autumn" (1967, 23). He adds, "The rich man who shows his wealth by spending recklessly is the man who wins prestige. . . . Consumption and destruction are virtually unlimited" (Mauss 1967, 35).

What is interesting is not only the superficial similarities between the reckless expenditures of contemporary wealthy Londoners and "premodern" Americans, but the explicit connection Mauss makes between consumption and destruction. It is through the very *consuming*, the actual shredding of wealth, the very destruction of plenty, that one gains prestige as one who is so wealthy he or she can *destroy* wealth and still retain it. Perhaps the court of Louis XIV and Nero's Rome were other milieus in which this kind of destructive consumption was elevated to a sort of maniacal art.

Yet perhaps it is the very economy of mass consumption that bears excess inside of itself. In a sense, this is an anti-reproductive regime, as conspicuous consumption proceeds to the point of destruction. In such an economy, every commodity must carry within it its own elimination, so that it can be destroyed and replaced, leading, ironically, to its own reproduction. Planned obsolescence is an economy of reproductive consumerism powered by commodities that deliberately cannot be reproduced and so must be repurchased as a whole, thus reproducing exchanges that exceed practicality, functionality, and need. Instead of a commodity reproducing its own functionality, commodities that cannot reproduce their own functionality reproduce new exchanges, thus reproducing buying and selling over and over again. And planned obsolescence is given many a stimulus by the quicksilver pace of change in fashion, fads, and trends, many of these anointed with fashionability by the constantly wielded prods of advertising and marketing.

Conclusion

Reproduction has now been explored in three distinct fashions: in the urban villages of China, as "circular" migrants and entrepreneurial "villagers" combine to form a strange hybrid of communism and capitalism to further their reproductive possibilities; in the case of the American middle class and its declining capacity to reproduce itself; and in the excessive reproduction of certain members of the ruling class. And reproduction as an essential aspect of the everyday has also been added to my schema, as reproduction on a daily basis is essential to the ongoing fabric of the everyday. The inclusion of reproduction also serves to bolster the quotidian aspect of the everyday: once conceived as those factors that reproduce life every day (eating, drinking, sleeping), the everyday becomes that much less elusive and that much more ordinary.

I want to add one more note on reproduction by way of Yi-Fu Tuan as well as Robert Sack. These two geographers make the point that it is the very processes of everyday behavior, repeatedly reproduced, that reproduce both places and whatever forces have inserted those behaviors in such places. As Sack puts it in *Homo Geographicus*: "Daily activities in space unconsciously reproduce the particular forces that have helped shape those routines" (1997, 159). Though Sack is quick to note that such routines and customs allow one not to have to worry about entirely reconstructing our world anew every day, they can also serve as props that allow patterns of injustice to continue. "If these very routines are somehow too limiting or unjust, spatial routinization of actions and conformity to rules . . . help perpetuate these limits and injustices, as well as making us unaware of them" (1997, 159). And everyday routine activities maintain as they reproduce and reproduce as they maintain the definitional parameters of the social relations taken for granted in those activities. "Each visit to a bank, store, airline, and vacation further instantiates the system. . . . Factories or offices . . . are places in which the social relations of workers and capitalists (and also the products they make) are produced and reproduced each and every moment" (Sack 1997, 76).

In "Rootedness versus Sense of Place," Tuan delineates ways in which places are not only created but also maintained, among which the everyday routine activities of "speech, gesture, and the making of things are the common means" (Tuan 1980, 6), so that reproduction of a place *as that particular place* depends on the reproduction of speech, gesture, and the making of things for its own ongoing maintenance *as that particular place*. Sack and Tuan's conceptions add a specifically geographical aspect to the notion of reproduction.

CHAPTER 5

Bringing in the Body

There is no position from which to view the everyday except via the body. In this chapter, I want to bring in the physical being, the perceptual network, the sensorial hub, for certainly without them there is no everyday. I accomplish this in three steps: first, through an examination of Merleau-Ponty and his reworking of Husserl's phenomenology, then through an excursus of the indexical body, and, finally, through a vindication of the sensorial as a systemic whole, or, in an inversion of this conception, through a repudiation of the singularity and autonomy of discrete and distinct senses.

Merleau-Ponty

"Carnal awareness is the Archimedean point of Merleau-Ponty's philosophy," states Richard Calverton McCleary in his preface to Merleau-Ponty's *Signs*, "for the body's presence to itself as both perceiving and perceived provides us with our fundamental knowledge of the consciousness in terms of which the self, the world, and other men are constituted" (1964, xvii). This installation of the core of Merleau-Ponty's philosophy into an Archimedean point situated in "carnal awareness" is a break from Husserl's form of phenomenology, which, though based in bodily experience, still gravitates to an ideal transcendental form of "egological" thought as its epicenter. Husserl's "reduction" is bracketed off from the world, removed from the body and its surroundings, as he posits a "phenomenological Ego" that "establishes himself as *'disinterested onlooker,'* above the naively interested Ego" (1999, 35; italics, Husserl). From this grand perch, somehow rendered "pure of all accompanying and expectant meanings on the observer's part," said observer will "create for itself a universe of absolute freedom from prejudice" from which to gaze upon the preselected aspect of the given, render judgments as to what is observed, and then make the phenomena "accessible by means of a new reflection, which, as transcendental, likewise demands the very same attitude of looking on *'disinterestedly'*—the Ego's sole remaining interest being to see and to describe adequately what he sees, purely as seen, as what is seen and seen in such and such a manner" (Husserl 1999, 35; italics, Husserl). On the other hand, Merleau-Ponty's "perception" is of the world, instilled and incorporated into the body and its surroundings: "Surely it is true that the new morphology of the visible acquired from the study of Merleau-Ponty's work does implicate a new conception of the ideal, which cannot be defined by opposition to the sensible" (Lingis 1968, li). And it cannot be defined by opposition to the sensible because it exists *within* the sensible, within the corporeal being, that is, within the body.

In the *Phenomenology of Perception*, what Merleau-Ponty accomplishes is to bring the question of the body and its inherent worldliness (its being-in-the-world) into philosophy: "We are involved in the world and we do not succeed in extricating ourselves from it in order to achieve consciousness of the world" (1962, 5). By establishing a "relation of continuity between what might be considered the 'interior' aspects of the subjects and the 'exteriority' of the world," Merleau-Ponty also establishes an ontology no longer "restricted by nature-consciousness (or body and mind) dichotomies" (Olkowski 1999, 1, 5). This "relation of continuity" is synonymous in Merleau-Ponty's system with the scope of the body or, better, the enfolding of the flesh. As Edward S. Casey explains this: "Continuity is guaranteed . . . by the lived body—which subtends all experience, even the most disjointed—and eventually by *flesh*, a term that brings body and world together in a literally convoluted but ultimately consistent manner: the 'flesh of the world' is a whole . . . in relation to which discontinuities can only be momentary fissures within its encompassing embrace" (2007, 68; italics, Casey).

So Merleau-Ponty brings the body back in after its long exile from Western thought. As Iris Marion Young puts this, "The unique contribution of . . . Merleau-Ponty . . . to the Western philosophical tradition has consisted in locating consciousness and subjectivity in the body itself. This move to situate subjectivity in the lived body jeopardizes dualistic metaphysics altogether. There remains no basis for preserving the mutual exclusivity of the categories subject and object, inner and outer, I and world" (Young, qtd. in Grosz 1999, 149). By demonstrating "that experience is always, necessarily embodied," Merleau-Ponty also demonstrates that "experience can only be understood between mind and body (or across them), in their lived conjunction, rather than, as Cartesianism implies, in their logical disjunction" (Grosz 1999, 149). And since there is nothing outside of experience, it follows that the mind-and-body conjunction (the flesh) must be the basis of any given ontology.

But what kind of body is this that Merleau-Ponty is describing? Male? Female? European? Asian? African? Healthy? Diseased? Heterosexual? Homosexual? Transgendered? In short, does it allow for difference? Or is it unified around a normative model of the body? Feminists, though appreciating Merleau-Ponty's invocation of the body as the field of ontology—"Merleau-Ponty, as one of the few more or less contemporary theorists committed to the primacy of experience, is thus in a unique position to help provide a depth and sophistication to feminist understandings, and uses, of experience in the task of political evaluation" (Grosz 1999, 148)—question the neutral postulation of such a body, for, by the default guaranteed by most theory prior to the new terms of discourse invoked by feminists, queer theorists, and others, such a body was white, male, and European/American. Therefore, many feminists, "while utilizing his work, nonetheless remain suspicious of the apparent sexual neutrality of his claims" and remain cautious about Merleau-Ponty, even to the point of dismissing the validity of his work as being solely restricted to a subject enwrapped in masculine experience. Yet, this limitation does not deter Grosz from stating that Merleau-Ponty's work can still be useful to feminists as "a new account of masculine modes of human being—a major revision of its scope and relevance" (Grosz 1999, 150, 151).

Judith Butler seems even more wary of Merleau-Ponty: she asserts that even though in the *Phenomenology of Perception* Merleau-Ponty ostensibly "offers certain significant arguments against naturalistic accounts of sexuality that are useful to any explicit political effort to refute restrictively normative views of sexuality," in the final analysis, "the potential openness of Merleau-Ponty's theory of sexuality is deceptive" (1989, 85, 86). This problem arises, according to Butler, because "despite efforts to the contrary, Merleau-Ponty offers descriptions of sexuality which turn out to contain tacit normative assumptions about the heterosexual character of sexuality" to the extent that he "conceptualizes the sexual relation between men and women on the model of master and slave" (1989, 86). So instead of offering up a universal or neutral conception of embodiment in which to enfold the duality of subject and object, Merleau-Ponty offers up a conception of embodiment based on a normative heterosexual voyeuristic "gaze" in which the male (Merleau-Ponty himself) places the female in a passive, oppressed position, thus replicating Hegel's master-slave paradigm. Butler admits that in *The Visible and the Invisible* Merleau-Ponty "suggests an ontology of the tactile [as opposed to his previous "social ontology of the look"], a description of sensual life which would emphasize the interworld, that shared domain of the flesh which resists categorization in terms of subjects and objects" (1989, 97). Yet the problem remains that Merleau-Ponty "fails to acknowledge the historicity of sexuality and of bodies. For a concrete description of lived experience, it seems crucial to ask *whose* sexuality and *whose* bodies are being described, for 'sexuality' and 'bodies' remain abstractions without first being situated in concrete social and cultural contexts" (Butler 1989, 98; italics, Butler). In effect, Butler is accusing Merleau-Ponty of a failure to practice self-reflectivity: he has not reflected upon his positionality as a white, privileged, European male vis-à-vis embodiment in general.

Still, Merleau-Ponty must be given credit for bringing forth the body as a possibility for others to extend, making his rendition of a particular form of flesh into universal flesh that includes the same as well as the other. Other writers, such as Maurita Harney in her "Merleau-Ponty, Ecology, and Biosemiotics," are willing to give Merleau-Ponty a lot more credit than this: "Merleau-Ponty . . . effectively shakes off the Cartesian-derived dualisms, most notably the oppositions of mind and matter, of subject and object, of culture and nature, and of human versus natural reality" (2007, 133). Whether Merleau-Ponty effectively did this is almost beside the point, at least for my purposes: what is necessary is that he has allowed the body to be brought front and center.

In Situ

"My" body is in the world only in reference to its own indexical position in both space and time. The *here* and *there*, the *near* and *far*, the *up* and *down*, the *left* and *right*, the *above* and *below*, and the *back* and *front* are all indexical placeholders that will vary from person to person depending on their respective positions. Your *near* may be my *far* and my *here* your *there*, for if it was also your *here* we'd be in exactly the same body. "Each person gets information about his or her body that differs from that obtained by any other person" (Gibson 1979, 115).

This may seem rather elementary and even simplistic, but it can quickly gain complexity to the point of utter knottiness. How is it that besides comprehending my indexical position in situ, I can comprehend someone else's *in situ*? And how is it that I can understand how your *there* could become my *here* and my *here* your *there*? Alfred Schutz provides some of the answer to this when he states, "In the everyday world in which both the I and the Thou turn up, not as transcendental but as psychophysical subjects, there corresponds to each stream of lived experience of the I a stream of subjective experience of the Thou. This, to be sure, refers back to my own stream of lived experience, just as does the body of the other person to my own body" (1967, 102). Now, while it is true that my comprehension of your indexical position "bears the mark of my own subjective Here and Now and not the mark of yours," I can imagine our places switching, given the possibilities of locomotion and duration (Schutz 1967, 105). Such a possibility "refers back to my own stream of lived experience," as Schutz says, but how is this so? Through experience, one gains the knowledge that, given the possibilities of movement and the passage of time, we can transpose ourselves into new positions. Again, elementary and simplistic. Yet this is precisely the kind of everyday assumption that makes ordinary communication and action possible. Without it, the everyday would collapse into silence and immobility.

This becomes trickier when Schutz's statement that one can reference *the body of the other person to one's own body* is considered. Here, Schutz seems to be imagining something like a perfect speech situation à la Searle or Habermas, in which intentionality and communication flow from one person to another without obstacles or misunderstandings. Merleau-Ponty comes up with a similar type of perfect communicative situation, though he puts it more abstrusely: "I project myself in the other and the other in me, [there is] projection-introjection, productivity of what I am doing in the other and of what he is doing in me, true communication through lateral practicing" (2010, 6). Once again, a sort of normative body has been posed, in which bodies are taken to be the same and completely aligned in that sameness.

It may seem that I have conflated the speech act situation with the indexical question, but the "context of meaning" bridges the gap between these two, as both speech acts and indexically oriented bodies can only proceed within such a context. The point I want to make here is that such contexts are not always aligned, one to another, not even in regard to their most basic elements. For instance, in *Getting Back into Place*, Edward S. Casey tells us: "In the case of human animals, the here-there and near-far are culturally (and linguistically) conditioned as well" (2009, 64). Given this, one can imagine an almost endless variety of conceptions of near, far, here, there, up, down, and so on, leading to communicative situations in which the shared context of meaning has been displaced to such a degree that communication is itself dismembered.

But it is even worse than that, as conceptions of the body itself are variable to a remarkable degree. Even the *same* body can experience itself in a great variety of ways, passing from one mean and mode of emplacement and displacement to another with almost quicksilver speed. How then to expect *different* bodies, each subtending *different* streams of lived experiences (to borrow Schutz's terminology) to comprehend one another in terms of respective indexical positionalities?

The problem only gets thornier if gender differences are tossed into the mix, which, via the modes of necessity and obligation, they must and should be. For instance, in her "Throwing Like a Girl: A Phenomenology of Feminine Body Comportment, Motility, and Spatiality," Iris Marion Young argues, "Feminine existence lives space as *enclosed* or confining," leading to a bodily experience of space and indexicality in which "the projection of [such] an enclosed space severs the continuity between a 'here' and a 'yonder'" (1990, 62, 63; italics, Young). So that, for "the female person who enacts the existence of women in patriarchal society" (Young 1990, 55), "the space of the 'yonder' is a space in which feminine existence projects possibilities in the sense of understanding that 'someone' could move within it, but not I. Thus the space of the 'yonder' exists for feminine existence, but only as that which she is looking into, rather than moving in" or moving toward (Young 1990, 63). Of course, the qualifier, for "the female person *who enacts* the existence of women in patriarchal society," leaves open the possibility that a female person may *not* enact such a mode of being, an option that becomes more and more feasible as emancipation increases and society becomes less and less patriarchal. But, "until female genitals and women's bodies are inscribed and lived (by the subject and by others) as a positivity, there will always remain paradoxes and upsetting implications for any notion of femininity," as well as for the possibilities of female indexical motility (Grosz 1994, 73–74). Still, even given a precisely egalitarian society with a sexually neutral power-sharing arrangement and a positive view embedded in each and every subject regarding the female body, there is no guarantee that gender differences will not obviate differences in bodily experience vis-à-vis indexicality. Even a matriarchal society still might lead (or especially might lead) to significantly different indexical gendered-inflected understandings of such things as the here and the yonder.

Yet we still manage to communicate, at least to some degree, and with at least some success. Generally, we still understand that, given enough time for locomotion, your *there* may become my *here*, your *near* my *far*, and so on. Even if these indexical transformations exist only as possibilities or as options restricted to someone of the opposite gender or of a different class, we still understand that they exist for someone, somewhere. How is this possible, given all the obstacles to such understanding, only a very few of which have been articulated? Why is Schutz at least generally correct when he states that even though "the environment I ascribe to you . . . bears the mark of my own subjective Here and Now," and that, furthermore, even though this "environment I ascribe to you . . . has already been interpreted from my own subjective standpoint" (1967, 105), "you" and "I" can still comprehend one another's respective indexical positions, viewing them as correlates of our own?

Perhaps some of the capacity to imagine your specific set of indexical coordinates in relation to my set, and, in particular, how your set could become my set can be explained by the recent discovery of the importance of mirror neurons in the functioning of the brain. Mirroring and imitation are not only necessary for learning language and motor skills but may also be a significant factor in the intersubjective perception of relative indexical positioning. "Your mirror system may help you be a better perceiver of others' actions," writes Lawrence D. Rosenblum in *See What I'm Saying: The Extraordinary Powers of Our Five Senses*; "it may do this by helping you to see others' behaviors based on how you might perform them yourself" (2010, 219). In other words, the pedagogical function of imitation by which we learn linguistic and motor skills plays a primary part in intersubjective transference, in which we imagine such things as your here becoming my here, your left my left, and so on.

But to foist off indexical coordination merely on the operations of each person's neurology would be a mistake. In the first place, conceived in such a way, as if encased in the carapace of one's own neurological network, every individual neurological system would be a kind of hermitically sealed shell, putting each set of mirror neurons in the odd position of having nothing to mirror. In the second, a holistic comprehension of intersubjective indexical coordination cannot be performed outside of an environment common to the subjects involved in such coordination. "If we want to understand the mind of an animal, we should look not only inward, to its physical, neurological constitution," writes philosopher Alva Noë in *Out of Our Heads: Why You Are Not Your Brain, and Other Lessons from the Biology of Consciousness*; "we also need to pay attention to the animal's manner of living, to the way it is wrapped up in its place" (2009, 42). Without a doubt, there are "internal correlates of consciousness. But there are external correlates of consciousness too" (Noë 2009, 42). And consciousness itself, argues Noë, does not exist except in its relationship to its environment. In effect, there is no such thing as a brain in a vat, isolated in its own absolutely self-referential cognitive universe. "To study mind, as with life itself, we need to keep the whole organism in its natural environmental setting in focus" (Noë 2009, 45). I return to this subject in much greater depth in the next chapter.

Gaining a sense of reciprocal spatiality is primary in the development of a child, and such a sense eventually leads to the coordination of relative indexical positioning as well. Referencing the French psychologist Henri Wallon, Grosz writes that during the phase of an infant's development of the sense of the spatial, "the child becomes not only able to distinguish the roles of agent and spectator (active and passive) but, more interesting, to play at both roles, giver and receiver, actor and audience, switching from one role to the other. This transitivism positions the child in a role of spatial reciprocity with the other, a space in which its position of the other is reciprocally defined by the position of the subject" (Grosz 1994, 48). Not only is this capacity to switch roles between the active agent and the passive spectator necessary for the development of language and motor skills, as already noted, but it is a prerequisite to the capacity of intersubjective comprehension in terms of the reciprocity of indexical coordination. If I do not have the ability to imagine myself in your position, then I cannot understand how your *right* could become my *right*.

All of this may be tied as well, one imagines, to the development of empathy, which was understood as an affect but has now been valorized by some scientists into a primary instinct. There can be no empathy without the capacity to comprehend myself as being in your position, thereby placing the senses of spatial reciprocity and intersubjective indexical coordination as primary constituents of empathetic understanding.

It needs to be made clear that a sense of spatial reciprocity is not some unlimited domain in which one's body is automatically interjected as a free-floating entity that can project itself into anyone else's positonality. In fact, there are many thinkers who believe that the capacity of intersubjective reciprocity is quite limited, as has already been seen in Young's reading of the female body as it is enacted in a patriarchal society. Clearly, the slave does not share a common reciprocal space with her master, in which the slave's place and the master's place are transferable units. As Hegel puts it in the famous section on Lordship and Bondage from the *Phenomenology of Spirit*: "one is the independent consciousness whose essential nature is to be for itself, the other is the dependent consciousness whose essential nature is simply to live or die for another. The former is lord, the other is bondsman" (1977, 115). However, this does not limit the slave (or the master, for that matter) from imagining the possibilities of intersubjective coordination. In fact, if the slave does not have the capacity to differentiate the master's indexical position from her own, she will not prove to be a very able performer of the tasks assigned to her, for what the slave must do "is really the action of the lord" and so must completely align herself in conformance to the lord's indexical situation (Hegel 1977, 116).

I suppose what I am attempting to clarify is the essential difference between a reciprocity of space defined as a domain of intersubjective congruence and one defined as a domain of intersubjective transference. In the former, physical coordination is maintained by a comprehension of respective indexical positioning; in the latter, bodily transference is presupposed as a physical possibility, regardless of one's status, gender, race, and so forth. As Young says, "The objects in visual space [including other bodies] do not stand in a fluid system of potentially alterable and interchangeable relations correlative to the body's various intentions and projected capacities. Rather, they too have their own *places* and are anchored in their immanence" (1990, 64–65; italics, Young).

If one returns to the subject of the everyday, one can quickly perceive that indexicality is key to mundane motility. As Casey points out, this is because our own sense of the indexical allows us to open up possibilities while closing off impossibilities, thus allowing for a feasible range of actions within limited horizons of space and time.

> Even as it acts to project a field of possible actions, my body closes down the prospect of unlimited choice. Hence it poses to itself constantly (even if often only implicitly) determinate choices between, say, going forward and retreating. Being in the center of things, my body can always move here *or* there, up *or* down, this way *or* that. This dimorphous structuring does not, of course, preclude still other possibilities, but it does bestow on a given field of possibilities a coherent set of routes. A spontaneous corporeal mapping or somatography arises in which, as on an actual map, meaningful alternate directions are available at each important juncture. (Casey 2009, 48; italics, Casey)

Casey articulates a component of the indexical that must be underlined. And this is that the indexical, being necessarily oriented around one's own body, as if radiating out from one's physical being, indexes one's body as the center of all things. Perhaps this is the natural and the only position in which the indexical can be located. However, such solipsistic physical coordination can also instigate the delusion that one truly is at the center of all things, even to the exclusion of all things other than one's own indexicality! But intersubjectivity and indexicality turn out to be intimately tied together: "We do not make contact with ourselves any more than we make contact with others" (Merleau-Ponty 2010, 134). This can be read as an axiomatic postulation of reciprocity between indexicality and intersubjectivity: there is not one without the other.

A fatal degree of indexical solipsism is evaded by the coordination of one's body with positions that are necessarily removed from one's body. Erwin Straus makes this clear in *The Primary World of Senses: A Vindication of Sensory Experience*: "I can only differentiate right and left as orientation marks in an object or with respect to my own body to the degree that I can *simultaneously* oversee directions in their manifoldness and mutual opposition" (1963, 390, 391; italics, Straus). Straus reports that mentally ill patients who cannot transcend their own subjectivity cannot differentiate left from right, up from down. Grosz reports on two different types of physic breakdowns of indexical positionality: hemiasomatognosia, "in which an organic disturbance, a cortical lesion, makes the subject unable to recognize one-half" of its own body, and alloesthesia, "generally manifested in a loss of the ability to distinguish between left and right sides . . . this occurs not only with respect to the patient's own body but also with respect to the bodies of others" (1994, 68). So we rely on our awareness of our own indexical position not only for orientation and locomotion but for our most basic mental functionality as well.

What J. J. Gibson refers to as "the persistence of the environment" leaves a kind of mysterious residue after it leaves our line of vision, thus encompassing the temporarily invisible into our indexical orientation.

> To say that one is aware of the environment behind one's head is to say that one is aware of the *persistence* of the environment. Things go out of sight and come into sight as the head turns in looking around, but they persist while out of sight. Whatever leaves the field as one turns to the right re-enters the field as one turns to the left. The structure that is deleted is later accreted; this is a reversible transition, and therefore the structure can be said to be *invariant* under the transition. (Gibson 1979, 208–9; italics, Gibson)

Now it is possible for the environment to vary to a slight degree while one is turning, but, at least in a general sense, Gibson is correct. This lingering persistence of the environment is what allows us to assume that indexical reciprocity is possible. It is also what allows us to assume the possibilities of coexistence and concurrence. Gibson describes such possibilities in the following passage: "Separated places and objects are perceived to coexist. This means that separated *events* at these places are perceived to be concurrent" (1979, 209; italics, Gibson). The reliance on the reciprocity between coexistence and concurrence, between space and time, is perhaps what leads to further assumptions of the existence of an objective world, for it can seem to be the "objective" world that guarantees the possibilities of reciprocally exchanged indexical coordinates.

However, the subjective perception must also be retained to a certain degree if an accurate sense of indexical coordination is to be maintained. As Straus indicates, "Right and left are always relative to the principal direction of the individual; the distinction between them depends upon the observer's standpoint and upon the direction of the person moving.... Only for me, and not for another who just passed this way, is that tree *there* before me" (Straus 1963, 391; italics, Straus). And so, in order to retain our indexical coordination, we must be geared into the world with equal measures transcendent objectivity and immanent subjectivity, transcending our own solipsistic orientation in order to perceive that your *there* could be my *here*, given the time to get *there*, while maintaining our own interior immanence as a subject with a discrete and distinct *here* and *there* circumambient to our body and in a distinct relationship to your body. No mean trick, that. But one that is very much taken for granted as a sort of everyday GPS somatic orientation system that moves us through and coordinates us to the rest of the world.

Another obstacle to a facile comprehension of indexicality is that every indication of the indexical (the here and there, the up and down, etc.) is multilocular. Casey points this out in *Getting Back into Place*: "But being here—what does that mean? Even if we grant that all being-here occurs through centration in the lived body ... there is not just one way to be here" (2009, 52). "Hereness," as Casey terms it, is a very labile, slippery term. "When the range of the here includes not just the place through which I am now moving (and which therefore reflects my immediate bodily movement) but all of the places to which I can effectively move, I experience a properly regional here. Examples include the entire house in which I reside ... a block on which I live, an entire neighborhood, even a country or state or nation" (Casey 2009, 53).

Here, then, is plastic, polymer, and polymorphous, and it can transform from the immediate vicinity of our body to the widely extended notion of the nation or even the globe or the universe. Indeed, we often extend our own *here* in a kind of prosthetic way as an extension of our body. This is probably most obvious when driving a car, as our physical being seems to amplify itself, the vehicle itself becoming an extension of our body. Patriotic fervor can also amplify our sense of the indexical here, as we identify our body with the geobody, the "body" of the nation and its various parts, the states, regions, or provinces. However, for my present purposes (which is to move the body into this discourse), there is no absolute need to resolve the issue of the lability of the *here*, constricting it so it only, say, circumscribes the immediate visual horizon of physical being. The indexical can simply be stipulated as a necessary ingredient in thinking about the body and let it be left at that.

Sensual Thought and the Body

What I want to do now is to bring the senses into my schema of the body. However, I want to do this in a particular fashion, for I want to advocate for a conception of the senses not as discrete organs that are viewed as distinct command-and-control centers of specific sensory capabilities but instead as a unified system in which each separate organ is intimately related to one another, to the point that they meld as one holistic system.

First off, breakthroughs in science seem to be corroborating a more "ecological" way of understanding how the perceptual system operates: "A significant shift in how perceptual phenomena are understood and evaluated has occurred" (Rosenblum 2010, xii, xiii). This "shift" is leading researchers to the conclusion that the sense organs are not autonomous, with each organ solely responsible for its own perceptual task. Instead, the sense organs seem to operate together, each organ combining to form what used to be considered percepts responding to single organs. In effect, given this new *perception* of perception, the ears see, the eyes touch, the skin tastes, the nose hears, and the tongue smells. It is also being acknowledged that the standard count of five discrete senses has limited our ability to comprehend not only the nuanced and holistic way in which the senses operate, but also the other sensory operations of the body that have been neglected due to a myopic focus on the five senses. Here, I am thinking of such "somatic sensations" as the sense of "kinesthesia (the sense of movement), proprioception (felt muscular position) and the vestibular system (sense of balance)" (Paterson 2009, 768). In fact, Mark Paterson writes that "current wisdom posits anything between eight and twenty-two" (2007, 20) different senses![1] While often slotted into the sensate domain of the tactile, these and other senses such as the sense of somatic pain and the sense of felt temperature (heat) "are difficult to resolve as distinct perceptions, evidenced by the fact that western medicine, psychology and social science has only relatively recently acknowledged them within the lexicon and there remains little consensus on the terminology" appropriate for these sensate operators or whether they should be classified as bona fide sensorial systems at all (Paterson 2009, 768).

In a study conducted at UCLA in 2011, psychologist Ladan Shams found that "interaction between sound and vision led to a significant improvement in detection of visual motion" (Wolpert and Menon 2011, 1). Shams compares her study, which involved three groups of participants tracking dots and sounds moving across a screen, with playing ping-pong: "Imagine you are playing ping-pong with a friend who serves the ball. You receive information about where and when the ball hit the table by both vision and hearing. . . . At least in regards to perception of moving objects, hearing and sight are deeply intertwined, to the degree that even when sound is completely irrelevant to the task, it still influences the way we see the world" (qtd. in Wolpert and Menon 2011, 2). Shams and the lead author of the study, Robyn Kim, report that "there are connections between the auditory and visual portions of the brain at the cognitive level. When the information from one sense is ambiguous, another sense can step in and clarify or ratify the perception" (qtd. in Wolpert and Menon 2011, 2).

However, rather than being a mere ratifying element or clarifying tool that only functions when possibilities of ambiguity arise, the "cooperation" between the senses seems to be much more foundational. Think of gazing at a concrete wall. Via memory, we know that its surface is rough; this haptic memory informs our sight so that our vision "feels" the wall as well as sees it. In fact, in *Sculpture and Enlivened Space: Aesthetics and History*, F. D. Martin uses precisely the example of a stone wall to "illustrate" the same concept of the intrasensory system. "We do not know about the surface, volume, and mass of these stones by sight alone but by sight synthesized with memories of tactual and kineaesthetic perceptions" (Martin 1981, 59, qtd. in Paterson 2007, 94). Husserl describes this as perceptions that are "suggested . . . by prior experience" (2002, 135). This process, in which sense memories can be transferred through our imagination to objects perceived in the present moment, pervades our perception of the everyday world. However, Mark Paterson in *The Senses of Touch: Haptics, Affects and Technologies*, asserts that intrasensory "teamwork" might be innate and biological, and so precede any instigation of memory. "Memory alone is not the intermediary of cross-modal [from one mode of perception to another] transfer. Instead, there are underlying encoding processes at a 'lower cognitive level' than memory, encoding experiences from different sensory subsystems" (2007, 45). While this could be true—that cross-modal transfer systems exist at a lower cognitive level than memory and therefore are prior to the instigation of specific memories—it seems to me that experience (and the uptake of experience through memory) must be one of the primary feeders of such transfer systems. We are not born with memories of the textures of concrete walls, for instance, but only gather them through experience. That such discrete memories may inform sense perception does not negate any prior existence of a cross-modal system; it merely makes them fully functional and renders them comprehensible to reason as well.

"Our research shows that your brain can detect . . . detailed information about sound-obstructing objects," writes Lawrence D. Rosenblum in *See What I'm Saying*, the very title of which is a reference to the fecundity of cross-modal sensorial systems (2010, 26). "It turns out that untrained, blind-folded subjects can hear the *shape* of small panels that obstruct loudspeakers" (Rosenblum 2010, 26; italics, Rosenblum).

Rosenblum, a professor of psychology at the University of California, Riverside, who has done work with blind painters who can "feel" the difference between colors and sightless mountain bikers who use echolocation devices "to follow trails and avoid large obstacles" (2010, 2), states that "new imaging technologies . . . show that your brain is much more ecumenical when it comes to your individual senses than once thought. . . . Areas of your brain once assumed to be dedicated to a single sense actually help out with multiple senses" (2010, xii). Oliver Sacks puts it this way: "There is increasing evidence from neuroscience for the extraordinarily rich interconnectedness and interactions of the sensory areas of the brain, and the difficulty, therefore, of saying that anything is purely visual or purely auditory, or purely anything" (Sacks 2003, 55, qtd. in Paterson 2007, 56). Such interconnectedness, such cross-modal perception, is typical, not unusual. For instance, the interdependence of vision and somatic motility is not incidental but foundational to the sense of vision. It could even be said that the rods and cones of the eyes, those basic instrumental receptors of vision, operate through motility, as they move and send signals when certain waves and particles of light touch them. Gregory Bateson explains this by stating that what is "typical of all sensory experience" is that our "sensory system . . . can only operate with *events*, which we call *changes*" (1980, 107; italics, Bateson). In regard to vision, Bateson admits:

> It is true that we think we can see the unchanging. We see what looks like the stationary, unmarked blackboard, not just the outlines of the spot. But the truth of the matter is that we continuously do with the eye what I was doing with my fingertip [sensing changes on the surface of a blackboard with his fingers]. The eyeball has a continuous tremor, called micronystagmus. The eyeball vibrates through a few seconds of arc and thereby causes the optical image on the retina to move relative to the rods and cones which are the sensitive end organs. (1980, 107)

And couldn't it also be said that the visual signals sent from the optic nerves to the brain are motile or depend, essentially, on motility? For how could they travel from the eye to the cortex if they could not move?

However, in *The Ecological Approach to Visual Perception*, J. J. Gibson downplays the entire notion of a motile conception of optic capabilities, stating that such a conception is a mechanical reduction of visual perception.

> It is not necessary to assume that *anything whatever* is transmitted along the optic nerve in the activity of perception. We need not believe that either an inverted picture or a set of messages is delivered to the brain. We can think of vision as a perceptual system, the brain being simply part of the system. The eye is also part of the system, since retinal inputs lead to ocular adjustments and then to altered retinal inputs, and so on. The process is circular, not a one-way transmission. . . . The eye is not a camera that forms and delivers an image, nor is the retina simply a keyboard that can be struck by fingers of light. (Gibson 1979, 61; italics, Gibson)

Gibson also asserts that it is the very conflation of the eye with the camera that has so confused and confounded those attempting to formulate theories of vision (1979, 64).

Furthermore, Gibson believes we compound our misconstrual of the visual system by basing our analysis on our understanding of human communication. "The concept of information with which we are most familiar is derived from our experience of communicating with other people and being communicated with, not from our experience of perceiving the environment directly" (1979, 62). Our tendency to comprehend systems by analogizing them to other systems leads to a miscomprehension of the visual system. Returning to Noë's point, somatic motility and visual perception are intimately united: as our bodies move, so our eyes see.

The interdependence or "teamwork" of the visual and the tactile has been recognized for quite some time. Indeed, in *A Treatise of Human Nature*, first published in 1740, David Hume states, "The idea of space is convey'd to the mind by two senses, the sight and touch; nor does any thing ever appear extended, that is not either visible or tangible" (1978, 38). Noë describes this interdependence of sight and body: "Blinking, turning the eye or head, or moving in relation to objects bring about characteristically eye-based sensory events" (2009, 60). Eye-based these "sensory events" may be, but *based* solely on the eyes alone they are not: there are no eyes in a vat. "What is remarkable is that the world can show up for visual consciousness—objects can show up with all manner of spatial and visible properties—thanks to our fluent appreciation of the ways that eye-related visual sensory stimulation depends on our movements" (Noë 2009, 60). This contrasts sharply with Husserl's claim that "the still body is the normal body," upon which we receive sensations that are in turn transposed and registered as perceptions by an activated brain (2002, 133). Indeed, the "normal" body is normally in motion, and the senses move with the body, as they are *in* the body and *with* the body, and so must move as the body moves.

But what is even more remarkable than Noë's claim that the world shows up for visual consciousness is that the world also shows up for those who are without sight, the blind. For the blind, space shows up despite the absence of visual stimuli; though the "collocation of senses of sight, touch and kinesthesia" are "a prerequisite for the production of sensory space, . . . it is nevertheless possible that another sense can become associated [with the creation of a sense of spatiality], one that offers the necessary spatial component such as audition" (Paterson 2007, 54). Indeed, the example of ping-pong, wherein the ball is located via sight *and* sound, as well as the example of echolocation serving as a navigational tool both for the motility of creatures such as bats as well as for blind mountain bikers, testifies to the efficacy of the association between auditory and spatial fluency. "Bats determine the size, location, density, and movement of prey such as fruit flies 100 feet away in a pitch-black cave by use of sonar, emitting frequencies of some 100,000 cycles per second, about five times what we can hear" (Sagan 2010, 23). The association between spatial features and audition "allows blindness [or, rather, allows blind animals, including blind human beings] to produce a non-visual, tactile and kinesthetic space independently from visual space, thereby reasserting the existence of true spatiality for the blind" (Paterson 2007, 54). The tactile sense is just as necessary for the sighted in constructing their sense of spatiality. A feeling for depth and the creation of spatial dimensionality are developed in sighted children through motile exploration and the associative pedagogy of the visual and the tactile working together as sensory "tutors" to instill in the child a knowledge of space.

The senses also never present anything as if the thing presented is possessed by separate senses and intended to be perceived solely through one discrete sense. Sartre describes this nicely in *Being and Nothingness*: "The lemon is extended throughout its qualities, and each of its qualities is extended throughout each of the others" (Sartre 1956, 186, qtd. in Merleau-Ponty 2004, 62–63). Analysis can separate these qualities in order to study them in the laboratory, and with much useful benefit. However, such a capability and its associated benefits should not deter one from recognizing that the senses (all eight to twenty-two of them!) operate *tout court*, not as separate units. And even in the laboratory, as already noted, scientists are now discovering that the reciprocity of the senses, not their autonomy, is what is foundational to perception.

As early as 1979, scientists such as J. J. Gibson were warning that experimental methods that treat vision as if it is autonomous should not be conflated with a reality that in actuality tends to holism rather than singularity. Gibson's research on "the ecological approach to visual perception" supports Merleau-Ponty's and Noë's thesis that vision is embodied and operates in tandem with movement. However, "looking around and getting around do not fit into the standard idea of what visual perception is" (Gibson 1979, 2). But looking around *and* getting around and, even more so, looking around *while* getting around are exactly what most animals, including humans, happen to do with their visual capacities. Gibson continues: "Note that if an animal has eyes at all it swivels its head around as it goes from place to place. . . . The evidence suggests that visual awareness is panoramic and does in fact persist during long acts of locomotion. . . . Ecological optics is required instead of classical optics" (1979, 2). Gibson notes that those who practice classical optics rely on evidence gained from studying subjects outside of the typical conditions in which the eye functions. Such methods inevitably lead to the conclusion that a disembodied, still eye subject to laboratory conditions is the proper unit of visual analysis even though vision operates in mobility through a changing and heterogeneous series of environments. Even the "still" eye moves as it blinks or switches its focus, shifting from one position to another. The conceit of a still eye is just that, a conceit.

Gibson attributes much of the confusion about visual perception to the discovery of perspective representation in the Renaissance by Leon Battista

Alberti and Filippo Brunelleschi: "to confuse pictorial perspective with natural perspective is to misconceive the problem of visual perception at the outset" (Gibson 1979, 70–71). Gibson also critiques "natural" perspective with its vanishing point as well; according to him: "It geometrizes the environment and thus oversimplifies it. The most serious limitation, however, is that natural perspective omits motion from consideration. The ambient optic array is treated as if its structure were frozen in time and as if the point of observation were motionless" (Gibson 1979, 70). Vision must be theorized, then, with at least two sets of motion in mind, one attributable to the point of observation (the eye) and the other to the ambient array of light (sunlight and its various reflections). To do otherwise is to ignore obvious empirical conditions.

Yet Gibson himself ignores another set of conditions that impacts vision as well as all the other senses, no matter how many senses we may tally as being actual, or if we conceive of the sensorial system as one unified structure or composed of discrete units; and that is the issue of cultural variance vis-à-vis the sensorium.

In her *World of Senses: Exploring the Senses in History and across Cultures*, Constance Classen states that "in the West we have two basic sensory paradigms for understanding the cosmology of other cultures," that of the visual paradigm adopted by the West and reflected in such terms as "world-view" and "perspective," and that of the oral-aural paradigm, which scholars typically assign to pre-literate societies (1993, 121). Yet Classen reports that there are completely other sensory modes by virtue of which societies "view" the world, thus demonstrating that sensory organization is not necessarily an either/or selection structured around *either* the visual *or* the oral-aural. For instance, "Thermal symbolism is widespread among the indigenous cultures of Latin America. Classificatory schemes based on concepts of heat and cold can be found from the southern Andes to northern Mexico . . . and indeed can vary widely from place to place" (Classen 1993, 122). For the "Tzotzil of the Chiapas highlands of Mexico," who are "descendents of the Maya," Classen indicates, "everything in the universe is thought to contain a different quantity of heat, or dynamic power" (1993, 122, 123). Such an emphasis on the thermal does not obviate the perception and appreciation of other sensate modes: "At the same time as Tzotzil rituals emphasize heat, the participants have their senses of smell and taste engaged by fragrance and food, their sense of hearing by music and speech, and their sense of sight by colourful decorations" (Classen 1993, 124).

For the Ongee, who live on Little Andaman Island in the Bay of Bengal, smell is the fundamental cosmic principle. "Odour is the source of personal identity and the reason for living in society.... By controlling odour, the Ongee control the cosmos" (Classen 1993, 126). And for the Desana of the Columbian Amazon, it is color that codes and controls the universe. This, then, is an order based on the visual, yet one that places its primary emphasis on "chromatic energies" (Classen 1993, 131).

Historical trends also play a role in the valorization of one or another of the senses. For instance, according to Classen, the olfactory sense was privileged in England throughout the Victorian era. This was reflected by the dissemination of fragrances through "pot-pourris, sachets, stationery, gloves and handkerchiefs," which "allowed for a discrete but intimate romantic discourse" (Classen 1993, 32). This came in handy: "In a society in which there were so many things which one was not allowed to see or say or touch, perfumes were charged with hidden meaning" (Classen 1993, 32). This olfactory regime slowly lost its power, first through an emphasis on the visual form of flower arrangements and then, much more resoundingly, with the negative association of the olfactory with the mustard gas of World War I. The valorization of the olfactory can be traced back to antiquity. "The heyday of the rose [and of fragrance in general] in the West was in the ancient world, where it was not only exalted in literature and mythology, but also used for a multitude of profane and sacred purposes" (Classen 1993, 17). Sight began to displace the olfactory as well as the auditory with the invention of the printing press and the discovery of linear perspective. The Enlightenment, with its emphasis on the "light" of reason, privileged sight to an even greater degree. As Constance Classen tells us, "Using the adjectives 'bright' and 'brilliant' to mean intelligent only came into vogue during the era of the Enlightenment, when the cultural importance of sight was on the rise" (2005, 5).

It seems to me that the close of the twentieth century and the beginning of the twenty-first have witnessed a gradual transformation in the conception of the senses. First, we are coming to the realization that the senses do not operate autonomously. For instance, note Paterson's claim that "haptic knowledges involve multiple relationships between the visual, the non-visual and the somatic senses" (2009, 781) as well as Rosenblum's assertions that the senses work together in a myriad of ways: "The growing evidence for our ability to use multiple senses for what until very recently were considered singe-sense functions ... supports the emerging notion that the brain is designed around multisensory input" (2010, xi).

Second, it appears that the tactile and the haptic in general are in the ascension, their status being driven forward by feminist and queer thinkers, frequently working in opposition to the predominance of the phallic gaze as well as by "objective" scientists who are beginning to realize that the tactile pervades our sensorium. The tactile is also in ascension due to the gradual change in computer and communication technologies from orientation to the visual (the screen) to orientation to the haptic (the *touch* screen, the mouse). "The role of touch in computing is becoming established and the various technologies of touch are becoming increasingly prevalent," writes Paterson in *The Senses of Touch*, published in 2007 (129).

In works such as Adam Greenfield's *Everyware* and Emile Aarts and Stefano Marzano's *The New Everyday: Views on Ambient Intelligence*, a haptically oriented world of ubiquitous computing (Greenfield's phrase) or ambient intelligence (the term of choice for Aarts and Marzano) is envisioned in which computing surfaces are embedded into nearly every available surface and are responsive to tactility, both at the proximate and the distal levels. In an essay titled "Towards a New Sensorality" in *The New Everyday*, Marion Verbücken writes, "The skin or surface of an object is the point of interface with users, and conveys the identity of the thing as well as containing its controls" (2011, 57). In such a world, the "skin" of the computer (or device) mirrors the skin of the user, creating a haptic interface between the two. Such technology does not nullify the visual (or any of the other senses, for that matter), yet it does lead to an "upgrading" of the tactile at the possible "expense" of the visual, perhaps creating a horizontal or holistic appreciation of the senses rather than the vertical hierarchy of the senses, with vision as the crown jewel of the sensorium. Designers of computers and other technologies are becoming more aware of the importance of the haptic "feel" of devices to their prospective consumers. "The sense of touch may prove to be one of our best salesmen, once we have relearned its secrets" (Sheldon and Arens 2007, 428). And one can see this concern with tactile compatibility reflected in marketing; for instance, on its website, Apple advertises its iPad in the following "haptic-friendly" manner: "With iPad, you use your fingers to do everything. And thanks to Multi-Touch technology, everything you do—surfing the web, typing email, reading books, swiping through photos, and switching between apps—is easier and a lot more fun. When your fingers touch the display, it senses them using electrical fields. Then it instantly transforms your taps, swipes, pinches, and flicks into lifelike actions. Just like that."[2]

Vaulting from the most contemporaneous technologies to some of the most ancient, the navigational techniques of South Sea Islanders can be recruited to the promotion of a haptic, corporeal turn. For the Puluwatans of Micronesia, though "acute visual powers enable them to read and follow the myriad nocturnal constellations with close discernment . . . they are by no means entirely dependent on vision, since difficult weather conditions can render unaided sight useless" (Casey 2009, 27). Indeed, Puluwatan navigators deploy *all* their senses when sailing between the islands of the South Pacific. Not only do they actively imagine the unseen reference island over the horizon; they sometimes deliberately shut out sight (even when it is unobstructed) in order to draw more fully on the other bodily senses. This is particularly true of estimating the direction and identity of ocean swells: to become fully sensitive to these deep currents, Puluwatans "steer by the feel of the waves under the canoe, not visually." One navigator reported that he would sometimes "retire to the hut on his canoe's outrigger platform, where he would lie down and without distraction readily direct the helmsman onto the proper course by analyzing the roll and pitch of the vessel as it corkscrewed over the waves" (Casey 2009, 27, 28).[3]

What I have attempted to do—via admittedly cursory explorations of Merleau-Ponty, indexicality, and the pansensorial mode of perception—is to bring the body in as a constitutive element of the everyday. Without it, no experience of the everyday is possible. We bring our body with us wherever we go. It underlies our experience, one could say. But accepting such a phrasing would give a false picture of the experiential situation, as if our body is something subsisting underneath us, somehow subtending experience as an adjunct thing. The body is *there*, right in the middle. It cannot be theorized away, à la Descartes of *The Meditations*. It persists, demanding recognition. And it is finally getting its due, as has already been noted. To add one more instance of this, in their *Philosophy in the Flesh: The Embodied Mind and Its Challenge to Western Thought*, George Lakoff and Mark Johnson critique "mainstream Western" philosophy's reliance on "transcendent reason, independent of human bodies or brains," arguing that a return to the body has the potential to destabilize the entire edifice of traditional philosophy (1999, 21). As Lakoff and Johnson put it, at the core of that tradition are the propositions that "human reason and human concepts are mind-, brain-, and body-free and characterize objective, external reality. If these tenets are false, the whole worldview [of traditional Cartesian philosophy] collapses" (1999, 22).

Given such a collapse and the introduction (or reintroduction) of an embodied mind into philosophy, Lakoff and Johnson assert that a radical re-arrangement of our metaphysical worldview is required. "Suppose human concepts and human reason are body- and brain-dependent. Suppose they are shaped as much by the body and brain as by reality. Then the body and brain are essential to our humanity. Moreover, our notion of what reality is changes. There is no reason whatever to believe that there is a disembodied reason" (1999, 22).

The everyday does not exist without the physical body, as the body is that which lives through the everyday. In the next chapter I project out from the body into the things of the world. This move is made with the realization that the world, the environs, the *umwelt*, that which surrounds us in the in-dexical here and how, is a constitutive part, not only of our bodies, but of our minds as well.

CHAPTER 6

Bring in Geography

I begin this chapter with an excerpt from Sartre's *Nausea* because the passage can serve as a kind of foreshadowing of the main conclusion I wish to draw.

> I turn left, I'm going to crawl into that hole down there, at the end of the row of gaslights: I am going to follow the Boulevard Noir as far as the Avenue Galvani. An icy wind blows from the hole: down there is nothing but stones and earth. Stones are hard and do not move.
>
> There is a tedious little stretch of street: on the pavement at the right a gaseous mass, grey with streams of smoke, makes a noise like rattling shells: the old railway station. Its presence has fertilized the first hundred yards of the Boulevard Noir—from the Boulevard de la Redoute to the Rue Paradis—has given birth there to a dozen streetlights and, side by side, four cafés, the Rendezvous des Cheminots and three others which languish all through the day but which light up in the evening and cast luminous rectangles on the street. (Sartre 1950, 37)

The idea I wish to emphasize here is that the mind does not stand alone in the world but is deeply implicated in the world. In effect, brain, mind, and world are one. Or, better, they cannot be taken as separated, which is perhaps a lesser claim but one much easier to ingest. In other words, what might be called the cognitive brain, the affective mind, and the *umwelt* (the indexically coordinated surrounding environment) work together, in an essential and co-constitutive manner. Given this, not only is there no brain in the vat, but there is also no mind in a vat and no world in a vat as well. Using material gleaned from the allied disciplines of cognitive psychology, biosemiotics, and the philosophy of mind, I argue that geography—conceived of in a broad sense as the environment in which we move and that simultaneously moves past us, those geographical factors we encounter on an everyday basis, including everything from geographic elements such as hills and creeks to components of the built environment such as buildings and roads—is a basic constituent of the brain. In other words, the brain *not only* functions in the world, it *is* the world, that world through which we move and to which we react on a moment-to-moment and place-to-place basis.

And so I am making a pitch for a kind of everyday neurogeography. I suppose this stance may be judged as opportunistic, a mere following of a fad, as one after another scholar jumps on the academic bandwagon of neuroscience. That's fine: let the accusations fly. I would not take the stance unless I thought it could stand. For I cannot ignore what seems to me to be at least a tacit recognition in much recent scholarship that the surrounding geography, the umwelt, the indexical geography, if you will, of our own respective everyday somatic positions, is a fundamental and constitutional part of the brain.

Starting with the Vibrancy of Matter

In her *Vibrant Matter: A Political Ecology of Things*, Jane Bennett writes:

> On a sunny Tuesday morning on 4 June in the grate over the storm drain to the Chesapeake Bay in front of Sam's Bagels on Cold Springs Lane in Baltimore, there was:
>
> > one large men's black plastic work glove
> > one dense mat of oak pollen
> > one unblemished dead rat
> > one white plastic bottle cap
> > one smooth stick of wood
>
> Glove, pollen, rat, cap, stick. As I encountered these items, they shimmied back and forth between debris and thing—between, on the one hand, stuff to ignore, except insofar as it betokened human activity (the workman's efforts, the litterer's toss, the rat-poisoner's success), and, on the other hand, stuff that commanded attention in its own right, as existents in excess of their association with human meanings, habits, or projects. (Bennett 2010, 4)

One thing that, at least at first glance, seems rather curious about Bennett's catalog is her negligence at not including herself. Her eyes picked these things out, her mind engaged with this particular concatenation of stuff, her brain sorted them into discrete units to be itemized and tucked away, later to be retrieved for use in *Vibrant Matter*, a book intended, among other things to (so says Bennett) "paint a positive ontology of vibrant matter, which stretches beyond received concepts of agency, action, and freedom sometimes to the breaking point" as well as "to dissipate the onto-theological binaries of life/matter, human/animal, will/determination, and organic/inorganic . . . to induce in human bodies an aesthetic-affective openness to material vitality" (2010, x).

It is not that she completely ignores her own complicity in this assemblage of things, for notice that she does write that "As *I* encountered these items, they shimmied back and forth between debris and thing," and so she has brought herself into this spool of stuff, invoking her own spatial-temporal proximity to the things she stumbles upon; yet she does not offer herself the same status that she offers to the glove, the pollen, the rat, the bottle cap, and the smooth stick of wood. They are numerated in the encounter while she is not. But this seems "natural," as she is not a thing, as the things listed most certainly are. Bennett's status as a human being gives her the necessary abilities and the requisite distance to gather the observed collocation of stuff together, postulate its existence as a theoretical bundle, and then return to it at a later time, the moment of composition.

Yet, later in the book, Bennett does invoke her own materiality as a thing among things, as the stuff of the body is pitched as congruent and contiguous with the stuff of the world. "Each human being is a heterogeneous compound of wonderfully vibrant, dangerously vibrant, matter" (Bennett 2010, 12). So, oddly enough, Bennett is claiming that, by identifying with our own thingliness, by fully *incorporating* ourselves, bodies transcend the state of being simple objects. In other words, we elevate ourselves by incarnating our selves—which, of course, is the opposite of the history of abasement to which the body has been "exposed." From being blamed for our own mortality to being castigated for its desire for physical contact to being shamed for its "lack" of thought, the body, at least in what is called the Western tradition, has served as a kind of handy scapegoat, a thing that must be hauled around with us until we die, a tiresome burden, a fardel-bearing drag, unwieldy and ungainly when not downright ugly or outright monstrous, an obstacle to clear and distinct thinking, as well as a profound impediment to any and every form of enlightenment. Is there any need to catalog the genealogy of the hysterical denigration of the body and its absolute bifurcation from the "higher" powers of the mind or spirit in the history of what is called "Western thought"? From St. Paul to Descartes to the valorization of quantification and rationality, the body has been a fugitive from both the temple and the academy.

This has led to a multitude of unfortunate consequences, most of which I have no time to examine. However, one that should be recorded is that this exile of the somatic has led to the odd and counterintuitive notion that the study of perception should be sectored off from bodily experience. Unmasking this strange tendency, Straus says, "Sensations belong to our mind. Because this mind is understood as extramundane, incorporeal consciousness, sensations, too, are taken out of the natural connection of experiencing and the corporeal being-in-the-world" (1963, 16). The notion that consciousness is solely incorporated within the brain and has no structural or logistical connection with the body and its environment may have been bolstered by "the fact that neuroscientists can produce sensations by direct stimulation of the brain," says Noë in *Action in Perception* (2004, 210). Noë rejects this idea on two counts: first, by snipping away at the logical deductions of such experimental results: "that it is possible to produce *some* experiences, it does not follow that it is possible to produce *all* experiences" (2004, 211; italics, Noë); and second, by reminding us that "the fact that one can manipulate experience by manipulating the brain" is not in any way, shape, or form tantamount to demonstrating "that the brain is sufficient for experience" (Noë 2004, 211). Furthermore, Noë concludes: "Experience is not caused by and realized in the brain, although it depends causally on the brain. Experience is realized in the active life of the skillful animal. A neuroscience of perceptual consciousness must be an enactive neuroscience—that is, a neuroscience of embodied activity, rather than a neuroscience of brain activity. . . . On the enactive approach, brain, body, and world work together to make consciousness happen" (2004, 227).

Let me repeat that, for emphasis: "Brain, body, and world work together to make consciousness happen." This may be a rather crude way of putting the thesis (which, by the way, is the main thesis of this chapter), as it suggests some sort of consciousness workshop, in which the elves of the brain join together with the elves of the body and the elves of the world to "make consciousness happen." However, I concur with Noë: it is precisely this triad, functioning in unison, that produces consciousness.

But no, that's rather clumsy as well. For my thesis is that it is precisely *not* as a triad that consciousness occurs, but as one thing that can, for purposes of speculation or experimentation, be abstracted into three parts but that is one biophysical whole "in reality." I would also substitute the terminology "cognitive mind, affective mind, and environment" for "brain, body, and world," but, otherwise, this is the very nub of the matter.

Yet it is not simply through the experiments of neuroscientists that this notion of a disembodied brain as the seat of all experience, as it were, arises, for this is one of the primary strands of our Cartesian legacy. In "Of the Senses in General," the Fourth Discourse in his *Discourse on Method, Optics, Geometry and Meteorology*, Descartes states: "We already know sufficiently well that it is the mind which senses, not the body; for we see that when the mind is distracted by an ecstasy or deep contemplation, the entire body remains without sensation, even though it is in contact with various objects. And we know that it is not properly because the mind is in the parts serving as organs to the exterior senses that it experiences sensation, but because it is the brain, where it exercises that faculty which is called common sense" (1965, 87). The parameters of much of the clinical experiments conducted by scientists (be they social, life, or earth scientists) hinges on just this body-mind bifurcation stipulated by Descartes.

Part of the problem of this approach is that it can lead to assumptions that fail to pan out when deductions based on experimental conditions are applied to the much more nuanced and variable conditions of the world. Dorion Sagan makes a nice case for this in her introduction to Jacob von Uexküll's *A Foray into the World of Animals and Humans, with A Theory of Meaning*: "Focusing on one aspect of the environment, as science does to isolate objects for study, presents an abstracted, truncated version of the elements under study that eventually comes back to haunt those who overgeneralize on the basis of an incomplete sample" (2010, 7–8). The fully incorporated human being is a being in relationship with his or her environment; such a construal defines the human being as co-constituent with his or her environment. There is no human being outside of those definitional parameters. Certainly, Descartes can state that "I examined closely what I was, and saw I could pretend that I had no body," but, as he readily admits: he is only pretending. For the pretense to be taken as actuality was one of the most monumental mistakes of the Western tradition (1965, 28).[1]

In his *Putting Science in Its Place: Geographies of Scientific Knowledge*, David N. Livingstone makes a related claim by arguing that the truths of science are contingent upon location. "The meaning of scientific theories is not stable; rather, it is mobile and varies from place to place. And that meaning takes shape in response to spatial forces at every level of analysis—from the macropolitical geography of national regions to the microsocial geography of local cultures" (Livingstone 2003, 4). Livingstone justifies this claim by first noting that science is an activity performed by humans and then by reminding us that "human activities always take place somewhere" (2003, 5). Next, this quite obvious claim is complicated by a "placing" of place within a wide array of contingent impingements: "where an individual, a social group, a state, or a subcontinent is located in material space is . . . highly significant" to scientific experiments and their outcomes (2003, 6). Livingstone acknowledges that the scientist "must also be sufficiently 'disarticulated' from their social environments to permit them to reshape the very settings they emerged from. Spaces both enable and constrain discourse" (2003, 7). Once experimental results are postulated and published, they can be met with all the contingencies susceptible to the vagaries of reception. "As ideas circulate, they undergo translation and transformation because people encounter representations differently in different circumstances. If theories must be understood in the context of the period and place they emerge from, their reception must also be temporally and spatially situated" (2003, 11–12).

Yet even experiments conducted in the same place can lead to disparate findings. "Every experimental physicist knows that very dissimilar things may happen under what appear to be precisely similar conditions" (Popper 1944, 133). Given this admission that dissimilar results can happen under precisely similar conditions, is it any wonder that *dissimilar* results can happen under *precisely dissimilar* conditions? In "The Poverty of Historicism," Karl Popper admits, "We can never be quite certain whether our laws are universally valid, or whether they hold only in a certain period (for example, perhaps, only in a period in which the universe expands) or only in a certain region (perhaps in a region of comparatively weak gravitational fields)" (1944, 136). Sagan's, Livingstone's, and Popper's arguments all go some way to supporting the thesis that bodies and their varied locations within various environments have profound effects, and that, in the final analysis, they must be taken into account when considering the "real-world" ramifications of the findings of scientific experiments.

An acknowledgment of "embodied consciousness" is an acknowledgment of reality (or the truth, I suppose), for *where else could consciousness be*? (Paterson 2007, 22). How could the mind *ever* be extramundane? Does it not *only* exist in the world? Even the most austere lab is of this world.

Bennett reminds us that some 90 percent of the human body is composed of bacteria, thus weakening the case for some sort of purely human subjectivity existing in absolute separation from the things "of the world." The porosity of the skin (the organ of touch not being an impregnable blockade against foreign "invasion" but a skein through which much "alien" matter finds passage) also testifies to the intimate conjunction between human beings and their environment, as atmospheric particles as well as a wide variety of other elements seep into the flesh despite whatever prophylactic effects we take against such "intrusions." But if we accept that there is such a connection, turning the stuff circumambient to the body into members of an assemblage that includes "us" as "actants" in this assemblage as well—so what? In other words, what are the consequences of human bodies being more than mere objects?

It seems that Bennett wants to draw two conclusions from such a connection, one generally political and the other generally ontological, although the latter may have political ramifications as well. On the first score, Bennett asks: "Why advocate the vitality of matter? Because my hunch is that the image of dead or thoroughly instrumentalized matter feeds human hubris and our earth-destroying fantasies of conquest and consumption" (2010, ix). By attempting "to bear witness to the vital materialities that flow through us and around us," or, in other words, by attempting to link "us" to the things of the world, Bennett hopes to forge an empathetic connection between human beings and things, which category includes worms, forests, and ponds as well as dead rats and bottle caps (2010, x). This concern (which may remind readers of the literature of environmental philosophy, including such writers as John Muir, Henry David Thoreau, Barry Lopez, Wendell Berry, Gary Snyder, and Sarah Whatmore) speaks to the need to counter a purely anthropocentric relationship *to* the things of the world with a biocentric relationship *with* the things of the world, things in the latter slot including both animate and inanimate material.

Now, while it may be true, as Noël Castree points out in a review of *Vibrant Matter*, that Bennett fails "to suggest both why and how our current societies could feasibly encounter the world in this way," and, furthermore, that Bennett demonstrates a "hesitation—or is it an inability—to spell-out why a new politics is not only necessary but achievable," for my purposes we can let Bennett slide on this score (2011, 5). For what I wish to glean from her book is not so much a handbook outlining the contours of the necessity and the feasibility of a new politics, but a first step to a basis for a neurogeography, which, as part and parcel of its ramificatory reverberations, will require not only a new politics but a new biology, a new sociology, and a new geography as well.

Bennett's other purpose is what she calls the re-enchantment of the world. This assumes the acceptance of two things: one, that the world was indeed enchanted at some previous point in time, and two, that this era of enchantment ended at another point in time. Weber enunciates the disenchantment of the world as it was folded within the "iron cage" of the bureaucratic order during the nineteenth century, that similar period of governmental consolidation and state centralization. Bennett's claim is that such a re-enchantment can occur if the material elements of life are "once again" assumed to possess vitality, the notion being "that moments of sensuous enchantment with the everyday world—with nature but also with commodities and other cultural products—might augment the motivational energy needed to move selves from the endorsement of ethical principles to the actual practice of ethical behaviors" (2010, xi).

But it is not a given that the world was ever enchanted in the first place. Such a notion smacks of the thesis of a golden age to which we *may, can, should,* or *must* return (depending on the modality in which we choose to approach such a return). Hobbes scoffs at such a notion while Rousseau hails it, believing we can return to such a state of affairs once we assent to the common will. However, assume that Weber is correct and that the world became disenchanted with the ascension of the bureaucratic state. This would mean that the Egyptian, Persian, Greek, and Roman Empires all existed within the era of enchantment, when matter was vital, imbued with agency, and humans and the various things of the world were united in some kind of intimate relationship. But who were these subjects of these empires who were so enchanted? Many were slaves. Indeed, it could be claimed that the Egyptians building the Great Pyramid "enjoyed" a vital connection with matter. In fact, it could easily be claimed that their connection with matter was so intimate that there was little to no differentiation between their physicality and the physicality of the stones upon which they labored.

The point I am trying to make here is that for a large portion of humanity, in antiquity as well as in this, the twenty-first century, transcendence from matter is not a choice. They are locked into matter, their muscles and bones so materialized that their very beings are smothered in material. For such people a "return" to the "vitality" of matter may translate into their present condition, a condition to which they may not aspire, indeed a condition into which they are locked by grim necessity. For those who have escaped from an overly intimate relationship with matter, re-enchantment may be a thing to struggle *against* rather than something to struggle *for*; such possibilities seem to have escaped Bennett's purview.

Nevertheless, Bennett's concept of the self as a thing in a web or a network of other things is a necessary first step in the formulation of the trinity (the three-in-one and the one-in-three) I am attempting to construct of cognitive brain, affective mind, and umwelten. For it is always in a circumambient indexical perception of stuff that we exist.

From Bennett's string of dead rat, bottle cap, work glove, stick, and oak pollen, all coming together outside Sam's Bagels one sunny June morning in Baltimore, I want to interpolate a similar string brought together by that great materialist and prototypical aficionado of every variety of ephemera, Walter Benjamin. First, it should be noted that Benjamin himself was the ur-excavator of everyday existence. *The Arcades Project* can be read as a document outlining a methodological approach for such an operation, with the phenomena explored by Benjamin ranging from "the literary and philosophical to the political, economic, and technological, with all sorts of intermediate relations" (Eiland and McLaughlin 1999, ix), a method encompassing everything from the "revised title page of [Charles] Meryon's" *Etchings of Paris* (Benjamin 1999, 414) to the "politics of finance under Napoleon III" (Benjamin 1999, 135) to the "true fairies of these arcades . . . the formerly world-famous Parisian dolls, which revolved on their musical socle and bore in their arms a doll-sized basket out of which, at the salutation of the minor chord, a lambkin poked its curious muzzle" (Benjamin 1999, 693). The history of the nineteenth century, Benjamin's object of research in *The Arcades Project*, "could be realized only indirectly, through 'cunning.' It was not the great men and celebrated events of traditional historiography but rather the 'refuse' and 'detritus' that was to be the object of study, and with the aid of methods more akin . . . to the methods of the nineteenth-century" (Eiland and McLaughlin 1999, ix).

But perhaps Benjamin himself puts it better when he states, "It [*The Arcades Project*] corresponds to a viewpoint according to which the course of the world is an endless series of facts congealed in the form of things" (1999, 14).

Once More, with Emphasis:
The Perceptual System

In the following excerpt from *Action in Perception*, Alva Noë cautiously approaches the conclusion that I have also come to, namely, that brain, body, and environment ("world") are intertwined all the way down.

> A reasonable bet, at this point, is that *some* experience, or some features of some experiences, are, as it were, exclusively neural in their causal basis, but that full-blown, mature human experience is not. This is supported by such facts as that, until now, neuroscientists have been able to produce only relatively primitive experiences by direct stimulation of the cortex. More important, it is just not clear, given the virtual character of perceptual content, why an internal representation would be any better than access to the world itself. (2004, 218–19; italics, Noë)

This tippy-toe method, fretted with numerous qualifications, gets bolder and bolder as Noë gets closer and closer to his commonsensical deduction: why are pictures accessed in the brain better than direct access to the world itself? No matter what conception one has concerning the origin of life—whether of the religious, intelligent design variety or of the functional, utilitarian variety—it simply makes biological sense to conclude that we are imbued with access to the world as well as access to our brain, the pathway between being the billions upon billions of our neuroreceptors. Noë brings in Wittgenstein before arriving at a simple yet startling conclusion: "This harkens back to Wittgenstein's idea that anything a picture in the head could do could be done by a picture held in the hand. We go a step further. Why do we need a picture at all? We are *in the world*" (2004, 219; italics, Noë). As he reiterates this point in *Out of Our Heads*, Noë states, "It is a mistake to think that vision is a process in the brain whereby the brain builds up a representation of the world around us" (2009, 185). As early as 1966, J. J. Gibson was saying pretty much the same thing: "The traditional theory of visual perception based on a retinal picture or image of each object is profoundly misleading" (1966, 54). It is misleading partially because, even though the visual system "registers some kinds of information that no other system can, such as the pigment color of surfaces," it also "combines with all the others and overlaps with all of them in registering objective facts" (Gibson 1966, 54, 53).

Where does this idea of a mind, not only cut off from its own body but also cut off from the world, come from? *Where* is the mind that operates *without* a body? And *where* is the mind that operates *outside* of the world? These are obviously rhetorical questions, to a certain extent, as I have spent much of this chapter tracking down the sources of this mistaken notion. Body and mind do not exist as separate entities, despite whatever appeal the idea may have. "The idea that consciousness is distinguished from physical being by being intrinsically directed toward or upon an object has so much intuitive appeal that there has not been and will not likely be an end of thinkers who are persuaded that there must simply be a way to avoid the idealistic and *solus ipse* consequences" of such a line of thought (Deely 2007, 6–7). Reasonable as this may sound, it seems to me that there is very little "intuitive appeal" to the idea that consciousness is essentially "intrinsically directed." Frankly, I don't get it. Intuition, in my opinion, runs in exactly the opposite direction. And so does science: "There is no empirical or philosophical justification for the idea that the brain alone is enough for consciousness" (Noë 2009, 181). This prejudice in favor of the brain as the sole seat of consciousness "naturally" works against the idea of the participation of the body in consciousness, leading to a misplaced epistemology or, perhaps better, an epistemology that completely lacks a place. Wendy Wheeler describes this when she states in *The Whole Creature: Complexity, Biosemiotics, and the Evolution of Culture*: "Conceptual knowledge is the product not of a disembodied mind, but, on the contrary . . . by an embodied and enworlded, indwelling mind" (2006, 61).

We are "enworlded" not only because we are in constant contact with "our" environmental surroundings (our environments as indexically oriented around us, our particular incarnations of the umwelten), but also because our minds have evolved as if they *are* connected with the environment, which they most assuredly are; otherwise, one is dead. The former notion is supported by Gibson's construal of the haptic system in *The Senses Considered as Perceptual Systems*. To the question, "How does a perceiver feel *what* he is touching instead of the cutaneous impressions and the bone postures as such?" (1966, 112; italics, Gibson), Gibson answers:

> In brief, the suggestion is that the joints yield geometrical information, that the skin yields contact information, and that in certain invariant combinations they yield information specifying the layout of external surfaces. At any one moment the orchestrated input from the joints . . . specifies a set of bone directions relative to the spine, to the head, and to the direction of gravity. The bones and the extremities are thus linked to the environment. At any one moment, the total input from the skin likewise specifies a pattern of contacts with touching surfaces, one of which is always the surface of support. The skin is thus also connected to the environment by this simultaneous pattern. (1966, 113)

This information from the joints, the bones, and the skin is "inputted" into the mind, thus connecting the mind directly to its environment, allowing us to traverse through and across gradients that are multifarious in texture as well as constantly shifting their shapes and their forms as we move through the world. Neurons in the brain do not operate in a vacuum; they operate in the world! "This is the key to our puzzle. What governs the character of our experience—what make experience the kind of experience it is—is not the neural activity in our brains on its own; it is, rather, our ongoing dynamic relation to objects, a relation that . . . clearly depends on our neural responsiveness to changes in our relations to things" (Noë 2009, 59). And where are these "things" and "objects" to which Noë refers? Of course, they are in the environment around us! Where else could they be? This conception there-fore implicates neural activity "happening" in the brain as happening in our indexical, geographical environs as well.

Noë claims that perception of *any* kind is absolutely dependent on bodily movement. "The world makes itself available to the perceiver through phys-ical movement and interaction. . . . All perception is touch-like in this way: Perceptual experience acquires content thanks to our possession of bodily skills" (2004, 1). Noë goes on to make the case that perception, even of the visual kind, relies on "sensorimotor knowledge," that is, the "practical grasp of the way sensory stimulation varies as the perceiver moves" (Noë 2004, 14). Our vision is coordinated with our movement, a self-adjusting mobile feed-back system gauging distance, perspective, and other indicators correlated to bodily awareness vis-à-vis locomotion. Noë emphasizes that this depen-dence on the body is not limited to the visual—the haptic also relies on the corporeal. He reports on a "prosthetic visual system known as the tactile-vision substitution system" that first receives visual information gathered by a "head-mounted camera" and then activates "an array of vibrators on the thigh of a blind person" (Noë 2004, 26). These images, translated into bodily sensations, allow the blind to move with "the experience of objects arrayed in three-dimensional space. She [the blind subject] is able to make judgments about the number, relative size, and position of objects in the environment" (Noë 2004, 26). Noë doesn't claim that this is a haptically oriented system: "Crucially, it is not a mode of perception by touch. . . . For touch is a way of perceiving by bringing things up against you, into contact with your skin" (Noë 2004, 26, 27). But it is "seeing (or quasi-seeing) without the deploy-ment of the parts of the body and brain normally dedicated to seeing, for ex-ample, the eyes and visual cortex. . . . Somatosensory neural activity realizes visual experiences" (Noë 2004, 27).

The logical step I want to make here is that mind, brain, and body do not constitute our whole being. There are no human beings who exist outside of their environment or externally to their history and geography. Just as there is no brain in a vat, there is no human being in a vat. We exist in environ-ments: the whole creature is an environmental one, our evolutionary history only traceable as it is coordinated with and shaped by environmental influ-ences, and our brains specifically constituted on an ongoing basis as an organ not apart from but in the world. In his conclusion to *Out of Our Heads*, Noë states, "The last twenty-five years have witnessed the gradual shaping of an embodied, situated approach to mind. . . . It is now clear, as it has not been before, that consciousness, like a work of improvisational music, is achieved in action, by us, thanks to our situation in and access to a world we know around us. We are in the world and of it" (2009, 186).

Now, if this is so, that is, if consciousness is formed due to "our situation in and access to a world we know around us" (in other words, our respective umwelten), that means that our indexical geographies are constitutive components of our minds. That is the subject of the next, and final section.

Geography and the Mind

First, let me stipulate what I am not proposing. I am not claiming that a sort of objectified universal geography is part of the mind. In other words, I am not making the claim that we are walking around with the world in our heads, the oceans and continents of the globe embedded within our skulls. What I am claiming, rather, is that formulating the brain as a stand-alone entity is a serious blunder, for it is a formulation bereft of contact with both the affective mind and the outside world (our indexically coordinated geographies). This is nothing other than a supposition that brain, mind, and world are *essentially* separated, and this supposition is akin to that contingency examined previously, that of place, space, and time. The formulation of these triads, each abstracted out from one another first linguistically and then as objects of research, has led to the conflation of the abstraction with the reality. The trinity of brain, mind, and world is a unified entity, each combination of the three oriented around our own biological and geographical indexicality. As Noë says, we are *in* and *of* the world, and there is no way around this metaphysical-ontological-existential state of affairs.

Neither am I claiming that there is some neutral subjectless subject that stands in as the universal object for our postulated collective mind. Such a neutral construct, if simply left blank, has a very strong tendency to be captured by whatever is presumed to be the normative subject in whatever era and place in which we happen to live. Noë is right when he states that we are of and in the world, but he misfires by not stipulating that this world is differentiated along a thousand degrees of otherness, and that the multifarious subjects of the world come to their worlds from a cornucopia of biogeographical angles. John Protevi analyzes precisely this problem in his *Political Affect: Connecting the Social and the Somatic* as he states that every subject is not only synchronously located but diachronically configured as well. What members of what Protevi terms the "embodied-embedded school" (that is, such thinkers as Noë and Bennett) tend to ignore is that in order "to understand fully the complex interplay of 'brain, body, and environment,' . . . we have to understand the diachronic, not just the synchronic, social environment. That means we have to study populations of subjects and the way access to skills training and cultural resources is differentially regulated along political lines" (2009, 29). And, while I agree with this assessment (think of the differentiation, for instance, between the educational resources available for the children of Watts as compared to those available for the children of Malibu and the difference that such a differentiation makes in the lives of these children), these differentiations are still inserted and experienced by subjects, no matter where or when they live, using minds that are embodied, embedded, and enworlded. As was demonstrated previously, the verticality of history is a constituent part of this schema and needs to be factored into the present analysis (or any other metaphysical-ontological-existential analysis, for that matter).

"The dispute about the actuality or non-actuality of thinking—thinking isolated from practice—is a purely *scholastic* question," states Marx in the "Theses on Feuerbach" (1994, 99; italics, Marx). On the face of it, it may seem odd to haul in Marx at this conjuncture; yet, what he is claiming here matches, I think, the point I am attempting to make even if he is positing it from a very different perspective. Thinking does not exist outside of practice, and practice is located in the world. And it is purely scholastic to suppose otherwise, that is, it is academic folderol, a confusion between life and a realm of pure thought only available through the methods made available within the artificial constraints of laboratory conditions, quantifiable analysis, or the truth tables of logical propositions. The materialism of Marx also belongs in this analysis. "My relationship to my surroundings is my consciousness," states Marx parenthetically in *The German Ideology*, aligning him snugly with the vital materialism of Bennett and the philosophy of mind of Noë, albeit again coming from a totally different direction wherein class struggle trumps everything else (1994, 117).

"And when you want to try to understand why scientists do what they do, especially biologists, one of the things you have to allow for is that they are trying to deny the reality of mind in a world which has mind" (Bateson 1991, 162). Gregory Bateson may seem a bit intellectually flabby here: what does he mean when he posits "a world which has mind"? However, Bateson unravels this, albeit obliquely, when he tells us that "if you are going to understand things and build explanatory systems, especially mental explanatory systems," that is, if one wants to comprehend systematically the way in which the mind works, then "you will want to have within the system you're talking about pathways that are relevant to that system" (1991, 165). This seems eminently commonsensical: pathways relevant to any system must be incorporated into any holistic analysis of such a system. Bateson continues: "That is, if you want to account for the route followed by a blind man, you will need to include the blind man's stick as a part of the determinant of his locomotion. So, if a mind is a system of pathways along which transforms of difference can be transmitted, mind obviously does not stop with the skin. It is also all the pathways outside the skin relevant to the phenomenon that you want to account for" (1991, 165).

Again, the same conception, coming from yet another direction: the mind *obviously* does not stop with the skin. What Bateson is claiming is that difference "makes a difference." In other words, differences trigger transformations in all sorts of ways: the difference in the direction of sunlight vis-à-vis the location of a plant causes heliotropic growth, the difference of hunger in an amoeba causes physical locomotion toward food. But how do such differences create "trains of effect," which, in turn, "become material of information, redundancy, pattern and so on" (Bateson 1980, 121)? In *Mind and Nature: A Necessary Unity*, Bateson tells us: "First, we have to note that any object, event, or difference in the so-called 'outside world' can become a source of information provided that it is incorporated into a circuit with an appropriate network of flexible material in which it can produce changes. In this sense, the solar eclipse, the print of the horse's hoof, the shape of the leaf, the eyespot on a peacock's feather—whatever it may be—can be incorporated into mind if it touches off such trains of consequence" (1980, 121–22).

Well, that's OK, as far as it goes. But as he continues, Bateson prompts more questions. "The difference itself does not *provide* the energy, it only *triggers* the expenditure of energy. We talk then about differences and *transforms of difference*. Obviously a neural impulse is a very different sort of thing from a difference in light or a difference in temperature. . . . When such differences are transformed in successive ways through the system, mind becomes a very complex network of pathways, some of them neural, some of them hormonal, some of them of other kinds, along which difference can be propagated and transformed" (Bateson 1991, 164–65; italics, Bateson). What the heck does Bateson mean by "transforms of difference" that are transmitted along systems of pathways? Well, what this seems to mean is that successions of differences in our indexically oriented surroundings create differences in our minds as we move through and between various environments. Our minds reflect these differences as they are engaged in a continual process of reconfiguration as we move through these differences.

The mind doesn't work as if it is divested of differences: it works through and in these differences, differences in light, temperature, heat, form, shape, color, and so on. The world of the self, our *umwelten*, as it were, is created in the nexus between our beings and our geography as we encounter that geography. If the world does not show up, if we are subjected to sensory deprivation of an extreme kind, the world will not show up, and there will be no world. "Take the mammalian visual system: If an infant were prevented from using his eyes . . . the lack of stimulation would prevent the establishment of the rich arbor of neural interconnectedness that is in fact necessary for mature vision" (Noë 2009, 94).

That idea that connection to the world is a constituent part of the mind is bolstered by the philosopher and semiologist John Deely when he reverses Cartesian ideational causality by stating, "The *of* in the idea refers not to the mind as producing the idea but to *that of which the idea makes the mind aware in producing it*. In other words, the *of* distinctive of the idea as such refers not backward to the idea's productive source as *my* idea or *your* idea, but outward to the objective term of an experience in principle suprasubjective and, insofar, accessible to others besides the one here and now forming the idea making that object present" (2005, 44; italics, Deely).

In effect, this is equivalent to Wittgenstein's claim that private languages do not exist: "Here I should like to say: a wheel that can be turned though nothing else turns with it, is not part of the mechanism" (2001, 81, remark 271). The primary mistake Descartes made was imagining that the cogito exists by itself. The thought must have its object (a thought is always *about* something), and in any final analysis, that object always leads back to the world. Every *cogito* must have its *cogitatum*; that is, "consciousness is always consciousness of something" (Merleau-Ponty 1962, 5). Merleau-Ponty is picking up from Husserl, who, in his Paris Lectures, delivered in 1928, states, despite Descartes being "France's greatest thinker" (Husserl is speaking in Paris, after all):

> Descartes commits this error, in the apparently insignificant yet fateful transformation of the ego to a substania cogitans, to an independent human animus, which then becomes the point of departure for conclusions by means of the prin-

ciple of causality. In short, this is the transformation which made Descartes the father of the rather absurd transcendental realism. We will keep aloof from all this if we remain true to radicalism in our self-examination and with it to the principle of pure intuition. We must regard nothing as veridical except the pure immediacy and givenness of the field of the ego cogito which the epoche has opened up to us. In other words, we must not make assertions about that which we do not ourselves see. In these matters Descartes was deficient. It so happens that he stands before the greatest of all discoveries—in a sense he has already made it—yet fails to see its true significance, that of transcendental subjectivity. (Husserl 1964, 9)

Without adjudicating the relative merits of transcendental subjectivity versus transcendental realism, one can, I believe, agree with Husserl that every *cogito* must have its *cogitatum*. In other words, Descartes mislaid the *cogitatum* in his enthusiasm for the *cogito*. It is not accurate to report "I think, therefore I am," for every thought is a thought *about something*, or a thought with an intention attached to it, as Husserl puts it.

And so, returning to the citation from Deely, the idea doesn't have its matrix in the idea itself, for its matrix is, in Deely's phrase, "outward to the objective term of an experience in principle suprasubjective." "By constituting ideas as that which the mind is directly aware of, Descartes, Locke, and those after them must posit *something else* on the basis of which the 'idea-objects' are presented. What this 'something else' would be—the mind itself precisely as acting, perhaps, as opposed to any results of such acting—they do not discuss in express detail" (Deely 1986, 16; italics, Deely). The idea (or for that matter, the brain or the mind) cannot have as its locus itself because that would consist of a doubling back upon itself and therefore be redundant and thus unnecessary.

"The relations of the subject to the objects of its surroundings, whatever the nature of these relations may be, play themselves out outside the subject, in the very place where we have to look for the perception marks. Perception signs are therefore always spatially bound" (Uexküll 2010, 54). And if perception signs are "always spatially bound," does that then mean that perception per se is bound to its environment and bound to it in a necessary way? Well, where else could perception be bound except to its indexical geography? Exclusively to the mind itself? The mind and all its percepts rolled up in the brain? Yet how could this be? The mind as a plenum from its very inception, its perceptions always already intact, just waiting for the right moment for them to unravel and reveal themselves? No, that simply cannot be.

Yet Uexküll's other notion articulated in this passage, that is, that "the re-

lations of the subject to the objects of its surroundings . . . play themselves
out *outside* the subject," seems off somehow or simply wrong. Those relations
do not play themselves out outside the subject but are constituted *both inside
and outside* the subject, as they exist simultaneously in the subject and in the
subject's object of thought as well, that is, its environs. One strong line, as in
Bennett's and Benjamin's thingly conceptions.

"When we apprehend of some natural or cultural entity—a tree, say, or a
flag—we are not aware directly of any mental state as such" (Deely 1986, 17).
For how could we be aware of a "mental state as such" without reference to
something outside of the mind? Here we might like to bring in Wittgenstein
from the *Philosophical Investigations*: "One would . . . like to say: existence
cannot be attributed to an element, for if it did not *exist*, one could not name
it and so one could say nothing at all of it" (2001, 21, remark 50; italics, Wittgenstein). This, of course, does not translate into the axiom that only things
encountered "in reality" can exist and therefore only they can be named:
for example, what about unicorns? But unicorns *do* exist, insofar as we have
imagined them, painted them, written about them, and so on. To continue
with Deely's line of thought: "Rather, we are aware of a tree or a flag, something an idea most definitely is not. At the same time, it is clear that a tree
we are looking at, in order to be present not merely in the physical environment but in our awareness as well, requires for this relative-to-an-observer
existence some factor within the observer on the basis of which the tree presumably existing in nature *also* exists as terminus of awareness (as item in
the Umwelt and intersection in a web of sign relations . . .)" (1986, 17; italics,
Deely).

As an item in the umwelt and as an intersection in a web of sign relations,
the tree (or the flag or the unicorn or what-have-you) exists then in a mutually constituted—constituted in the moment of perception mutually by the
object (the tree), the subject (the perceiver), and the world (the umwelt)—
biosemiotoic relationship. This triadic constitution is a unified action; one
thing. It does not exist as three separate components joined together to form
a percept in the human brain. It is *absolutely* one thing, but verb-like rather
than noun-like, perpetually in motion, and does not exist except as suspended in that one moment between object, subject, and umwelt.

"It *would* be astonishing to learn that you are *not* your brain," states Alva
Noë in *Out of Our Heads* (2009, 7; italics, Noë). "All the more so to be told
that the brain is not the thing inside of you that makes you conscious be-

cause, in fact, there is no thing inside of you that makes you conscious. . . . To understand consciousness in humans and animals, we must look not inward, into the recesses of our insides; rather, we need to look at the ways in which each of us, as a whole animal, carries on the processes of living in and with and in response to the world around us" (Noë 2009, 7). If the mind is part and parcel of the process of living, which it surely must be, and if, as Noë claims, the processes of living are carried on "in and with and in response to the world around us," which they also surely must be, then that means that geography, as a stand-in for "the world around us," and the mind are intertwined to the point of unification, or at least to the point of inextricability, which amounts to pretty much the same thing. Such a construal places geography at the epicenter of neuroscience, as the world around us, our indexically coordinated geographies, must be accounted for in any viable conception of the brain or the mind.[2]

Now I want to return to Protevi's admonition that those investigating the "new cognition" not ignore social and political effects and their influence on the development of the mind. If Marx is correct when he claims "My relationship to my surroundings *is* my consciousness," then every aspect of one's surroundings must be incorporated into a study of what constitutes consciousness (1994, 117; italics, mine). An investigation of consciousness that leaves out factors—such as access to shelter and food, the role of gender in society, racial or ethnic variations in access to employment and educational opportunities—is an investigation that has not fully encompassed the variety of factors that are encountered by an embodied, embedded, and enworlded consciousness.

Keeping that caveat in mind, it seems to me that a conception of cognition that melds brain, mind, body, and world is one in which a conception of a more optimistic politics can be formulated. For, once we have assumed that we are creatures whose minds are constitutively connected to our bodies as well as to our respective environments, then a more responsible, progressive, and revolutionary politics becomes possible. Divorced from our bodies and the world (our geographies), we can easily be manipulated into believing that we are discrete and distinct self-interested entities cut off from one another. Connected to our bodies and the world, we become much more prone to perceive the world as a collectivized whole, our flesh one flesh, our ground one ground, yours *and* mine, in other words, *ours*.

It should also be kept in mind that a new mode of everyday consciousness has also been postulated here, one that incorporates the triad of mind, body, and world. The recognition of this mode of consciousness, a mode that we always already possess, is a requisite step to a clarified understanding of the everyday. Without it, human beings are disconnected from one another and from the world. With it, such connections are possible.

CONCLUSION

While I think it is fair to say that a considerable amount of ground has been covered since we commenced, it is just as fair to ask if I am any closer to capturing that elusive creature, the everyday.

While it never will be possible to demonstrate that the everyday has been captured, it does seem to me that the everyday becomes much less elusive when geography is inserted into the equation. Instead of starting from a theoretical nowhere (which, in effect, is tantamount to a theoretical everywhere), if we actually place our investigation *somewhere*, our odds for success (howsoever defined) improve significantly. Once the everyday is grounded in a place—the urban villages of China, Dreamland in Egypt, a subterranean mansion in London—the mystery of the everyday diminishes considerably because it has coordinates and is inserted into a concrete place rather than an abstract notion, which, being abstract, will always be elusive.

I hope that the work done here has at least equalized the field upon which we hunt down the everyday, narrowing the everyday's chance for continual escape while broadening our chances to nab the thing. For it seems to me that if we take our tools and apply them, always remembering that they can only be implemented in free-form fashion in no necessary sequential order and with no structural prerequisite for the locus of their application, then they can be of invaluable assistance in formulating the workings of the everyday. What I would also like to suggest is that bringing in geography might help many an elusive endeavor; however, that suggestion may be beyond the scope of my immediate concerns and will have to wait for its articulation until a later endeavor.

However, before hailing the efficacy of these tools and the triumph of this project, perhaps I should review my itinerary, that is, recall the route I have traversed in this study and thereby also renew the reader's acquaintanceship with my set of instruments. I started with Goffman and the situation, a loose-limbed thing as "theorized" by Goffman, and a thing also restricted by its Anglo-American matrix. I then compared and contrasted Goffman's situation with Foucault's milieu, and I concluded that Hacking's construal of a top-down Foucault being complemented by a bottom-up Goffman is a bit simplistic, if not simply mistaken. But Goffman's microsociology, his analysis of the minutia of the situation, should never be discounted: his powers of observation are sound, and his terminology and conceptual frameworks, variable as they may be, are extremely useful in analyzing any and every situation, no matter what its location. And Foucault is extremely useful as well, especially his conception of security and risk and how these concepts play out in the milieu.

I then battened down the situation by laminating it with time, space, and place while also arguing for a timespaceplace "thing" as a replacement for the scientifically antiquated notion of the discrete entities of time, space, and place or the more nuanced conceptions of spacetime or timespace. Next, I added the vertical element of history, as no present-day intersection of time-space-place exists without a diachronic axis running through it, deepening it with the residue of time. Next was reproduction, the necessities pertaining to daily reconstitution, and again I moved the geographical coordinates wider as I cited the housing situation as it is being played out in China, London, and Egypt. I brought in the somatic at that point, as no one has ever set foot in an everyday place or maneuvered their way through an everyday situation without bringing their body along. Finally, I conjoined body, mind, and environment, and raised the question if this schema might be the proper mental-somatic-geographic method by which to not only construe the mind but also approach the everyday.

And so I might envision using these tools in the following way. Taking up the situation of Goffman and the milieu of Foucault, I could make a sighting of that which appears before me, fixing it as a case that can be calibrated through this mobile analytical toolkit. I then remember that this is a situational milieu in which timespaceplace are operational as one indivisible whole, and that the case also has its own vertical historicity as well as its own horizontal spatiality. Bring in a reproductive analysis as it is reckoned how the present case may be satisfying the requirements necessary for everyday reproduction. Last, the body and things themselves are added, pulling these elements into the analysis to form a holistic diagram of the situation at hand. If this methodical sequence doesn't tell the whole story, the tools can be applied in sets of alternative sequences until it does.

This book is obviously a work of synthesis. I have ranged across various fields, including philosophy, physics, political economics, sociology, anthropology, history, neuroscience, and geography. Being by profession a teacher of geography, I would like to say something about geography as a synthesizing discipline. Long troubled by a disciplinary identity crisis, geography has had trouble deciding what its task is and what its role should be within academia (Sullivan 2011, 151). To many scholars, both within and without the discipline, this has seemed to be a source of great concern. I, however, regard it as geography's greatest strength in that such a lack of any definitive definition allows geography a wide scope upon which to ply its wares. If we assume a rather simple definition of geography, to wit: "The study of the surface of the Earth and all the life thereon," such a definition allows for and even calls for great latitude, if you will, both in subject matter and in methodology. We can, following this definition, allow for such studies as the variables of beach erosion and the elusiveness of the everyday. And works of scholarship from an expansive ambit, from Marx to Einstein and from Hesiod to Foucault, can be recruited. However, it may appear that during the course of this study I cast the net too wide, almost as if I intended the net to cover every ocean and every sea. But *The Geography of the Everyday* was from its very inception intended to be comprehensive; therefore, an almost circumpolar tossing of the net was required.

Now, a final note of vindication. *Where* we are must be included in any analysis. Just as there is no brain in a vat and no body in a vat, there is no analysis in a vat and no book in a vat. Geography matters. Without it, there is no everyday. With it, the everyday has a place and appears.

Chapter 1. Starting with Goffman and Ending with Foucault

1. Goffman quotes Ivan Belknap, *Human Problems of a State Mental Hospital* (New York: McGraw-Hill, 1956), 194.

Chapter 2. The SpaceTimePlace "Thing"

1. In no way do I intend to insinuate here that Hägerstrand has poached upon Goffman, committing an act of plagiarism, that is. Once the concept is in hand, these two examples nearly jump out at one, especially given the similar time frames within which the two men were working.

2. Michael Curry first brought this book and its geographic ramifications to my attention.

3. Gaukroger continues with Descartes's argument, which may be of interest: "It is important to note here that [in Descartes's schema] both the contents of the intellect *and* the contents of the world must *both* be represented in the imagination" (1995, 171; italics, Gaukroger).

Chapter 4. What Marx Brought in from the Cold: Reproduction

1. In a footnote, Nugent and Robinson concede that this percentage may be as low as 25 percent.

2. Engels cites Hawkins from the *Factories Enquiry Commission: Second Report of the Central Board . . .* (London: House of Commons, 1833), 3.

3. Harvey provides no sourcing for the citation of Budd.

4. Fliegelman cites Robert Blair St. George, "Artifacts of Regional Consciousness in the Connecticut River Valley, 1700–1780," in Robert Blair St. George, ed., *Material Life in America, 1600–1860* (Boston: Northeastern University Press, 1988), 349.

5. Barber cites a letter from Hoover to Roy A. Young, governor of the Federal Reserve Board, 24 March 1930, Presidential Subject Files, Herbert Hoover Presidential Library.

6. Watts is quoting from Karl Marx and Frederick Engels, *The Communist Manifesto* (London: Verso, 1998), 39.

7. Ghannam is quoting from *Shopping Centers Today* (a publication of the International Council of Shopping Centers), 1 May 1999, www.icse.org.

Chapter 5. Bringing in the Body

1. Paterson references B. Durie, "Doors of Perception," *New Scientist* 185 (29 January 2009): 33–36.

2. Quotation taken from promotion copy about the iPad on the 2012 Apple website, apple.com/ipad/features/.

3. Casey references David Lewis, *We, the Navigators* (Honolulu: University of Hawaii Press, 1972), 87, for the first two citations and Thomas Galdwin, *East Is a Big Bird* (Cambridge, MA: Harvard University Press, 1970), 171, for the last citation. Casey also adds that Lewis states that "holding courses by swells seems always to be a matter more of feel than sight" (87). For more on Micronesian navigation, see Norman Thrower, *Maps & Civilization: Cartography in Culture and Society* (Chicago: University of Chicago Press, 2007).

Chapter 6. Bring in Geography

1. Avicenna's "Floating Man" thesis smacks of the same pretense.

2. I realize I have been slipping between the terms "brain" and "mind." Rather than writing a tome justifying this, let us simply stipulate that since no one has ever successfully separated brain from mind or mind from brain, the alternation stands.

BIBLIOGRAPHY

Aarts, E. H. L., and S. Marzano. 2003. *The New Everyday: Views on Ambient Intelligence*. Rotterdam: 010 Publishers.

Agnew, J. 2005. Space: Place. In *Spaces of Geographical Thought: Deconstructing Human Geography's Binaries*, ed. P. Cloke and R. Johnston, 81–96. London: Sage.

Ahearne, J. 1995. *Michel de Certeau: Interpretation and Its Other*. Stanford, CA: Stanford University Press.

Ajayi, J. F. A. 1991. Nigeria, in *N–O*, vol. 14 of *The World Book Encyclopedia*, 414–23. Chicago: World Book.

Arendt, H. 1983. Bertolt Brecht: 1898–1956. In *Men in Dark Times*. San Diego: Harcourt Brace.

Aristotle. 1955. *Physics*. Revised text by W. D. Ross. London: Oxford University Press.

Barber, W. J. 1985. *From New Era to New Deal: Herbert Hoover, the Economists, and American Economic Policy, 1921–1933*. Cambridge, UK: Cambridge University Press.

Barry, A. 1996. Line of communication and spaces of rule. In *Foucault and Political Reason: Liberalism, Neo-Liberalism and Rationalities of Government*, ed. A. Barry, T. Osborne, and N. Rose, 123–42. Chicago: University of Chicago Press.

Bateson, G. 1980. *Mind and Nature: A Necessary Unity*. Toronto: Bantam Books.

———. 1991. *A Sacred Unity: Further Steps to an Ecology of Mind*. Ed. R. E. Donaldson. New York: HarperCollins.

Benjamin, W. 1999. *The Arcades Project*. Trans. H. Eiland and K. McLaughlin. Cambridge, MA: Belknap Press of Harvard University Press.

Bennett, J. 2005. The agency of assemblages and the North American blackout. *Public Culture* 17 (3): 445–65.

———. 2010. *Vibrant Matter: A Political Ecology of Things*. Durham, NC: Duke University Press.

Bentley, E. 1971. *Thirty Years of Treason: Excerpts from the Hearings before the House Committee on Un-American Activities, 1938–1968*. New York: Viking Press.

Blanchot, M. 1987. Everyday speech. Trans. S. Hanson. In "Everyday Life," ed. A. Kaplan and K. Ross, *Yale French Studies* 73, 12–20. New Haven, CT: Yale University Press.

Bluestone, B., and B. Harrison. 1982. *The Deindustrialization of America: Plant Closings, Community Abandonment, and the Dismantling of Basic Industry*. New York: Basic Books.

Bonta, M., and J. Protevi. 2004. *Deleuze and Geophilosophy: A Guide and Glossary*. Edinburgh: Edinburgh University Press.

Bourdieu, P. 1990. *The Logic of Practice*. Trans. R. Nice. Stanford, CA: Stanford University Press.

Bourdieu, P., et al. 1990. *Photography: A Middle-Brow Art*. Trans. Shaun Whiteside. Stanford, CA: Stanford University Press.

Brazil's Embraer spreads wings in China. 2011. United Press International (UPI). 22 August. http://www.upi.com/Business_News/Security-Industry/2011/08/22/Brazils-Embraer-spreads-wings-in-China/UPI-18741314048888/.

Brecht, B. 1975. *Schweyk in the Second World War*. In *Brecht: Collected Plays, Volume 7*, ed. R. Manheim and J. Willett. Trans. M. Knight and J. Fabry. New York: Vintage Books.

Brenner, R. 2002. *The Boom and the Bubble: The US in the World Economy*. New York: Verso.

———. 2006. *The Economics of Global Turbulence: The Advanced Capitalist Economies from Long Boom to Long Downturn, 1945–2005*. London: Verso.

Burns, T. 1992. *Erving Goffman*. London: Routledge.

Butler, J. 1989. Sexual ideology and phenomenological description: A feminist critique of Merleau-Ponty's *Phenomenology of Perception*. In *The Thinking Muse: Feminism and Modern French Philosophy*, ed. J. Allen and I. M. Young, 85–100. Bloomington: Indiana University Press.

Buttimer, A. 1971. *Society and Milieu in the French Geographic Tradition*. Monograph series, no. 6. Chicago: Rand McNally for the Association of American Geographers.

Capek, M. 1961. *The Philosophical Impact of Contemporary Physics*. Princeton: D. Van Nostrand.

Casey, E. S. 1997. *The Fate of Place: A Philosophical History*. Berkeley: University of California Press.

———. 2007. Borders and boundaries: Edging into the environment. In *Merleau-Ponty and Environmental Philosophy: Dwelling on the Landscapes of Thought*, ed. S. L. Cataldi and W. S. Hamrick, 67–92. Albany: State University of New York Press.

———. 2009. *Getting Back into Place: Toward a Renewed Understanding of the Place-World*. 2nd ed. Bloomington: Indiana University Press.

Castree, N. 2011. Review of *Vibrant Matter: A Political Ecology of Things*, by Jane Bennett. *Society & Space*, 14 September.

Chu, H. 2011. Emerging economies faring better in downturn. *Los Angeles Times*, 21 August.

Classen, C. 1993. *World of Senses: Exploring the Senses in History and across Cultures*. London: Routledge.

———. 2005. Fingerprints: Writing about touch. In *The Book of Touch*, ed. C. Classen, 1–9. Oxford: Berg.

Collier, S. J., and A. Ong. 2005. Global assemblages, anthropological problems. In *Global Assemblages: Technology, Politics, and Ethics as Anthropological Problems*, ed. A. Ong and S. J. Collier, 3–21. Malden, MA: Blackwell.

Conley, T. 1992. Michel de Certeau and the textual icon. *Diacritics* 22 (2): 38–48.

Cossery, A. 1949. *The House of Certain Death*. Trans. S. B. Kaiser. New York: New Directions Books.

Craig, W. L. 2001. *Time and the Metaphysics of Relativity*. Dordrecht: Kluwer Academic.

Crang, M. 2005. Time: Space. In *Spaces of Geographical Thought: Deconstructing Human Geography's Binaries*, ed. P. Cloke and R. Johnston, 199–220. London: Sage.

Davies, K. 2001. Responsibility and daily life: Reflections over timespace. In *Timespace: Geographies of Temporality*, ed. J. May and N. Thrift, 133–48. London: Routledge.

Davis, M. 2001. *Magical Urbanism: Latinos Reinvent the U.S. City*. London: Verso.

de Certeau, M. 1984. *The Practice of Everyday Life*. Trans. S. Rendall. Berkeley: University of California Press.

———. 1988. *The Writing of History*. Trans. T. Conley. New York: Columbia University Press.

Deely, J. 1986. The coalescence of semiotic consciousness. In *Frontiers in Semiotics*, ed. J. Deely, B. Williams, and F. E. Kruse, 5–34. Bloomington: Indiana University Press.

———. 2005. *Basics of Semiotics*. Tartu, Estonia: Tartu University Press.

———. 2007. *Intentionality and Semiotics: A Story of Mutual Fecundation*. Scranton, PA: University of Scranton Press.

Deleuze, G., and F. Guattari. 1987. *A Thousand Plateaus: Capitalism and Schizophrenia*. Trans. B. Massumi. London: Athlone Press.

Descartes, R. 1965. *Discourse on Method, Optics, Geometry, and Meteorology*. Trans. P. J. Olscamp. Indianapolis: Bobbs-Merrill.

———. 1974–86. *Oeuvres de Descartes*. Ed. C. Adam and P. Tannery. 2nd ed. 11 vols. Paris: J. Vrin.

Dionne, E. J., Jr. 2011. The last Labor Day? *Washington Post*, 4 September.

Eickelman, D. F. 2009. Re-reading Bourdieu on Kabylia in the twenty-first century. In *Bourdieu in Algeria: Colonial Politics, Ethnographic Practices, Theoretical Developments*, ed. J. E. Goodman and P. A. Silverstein, 255–67. Lincoln: University of Nebraska Press.

Eiland, H., and K. McLaughlin. 1999. Translators' foreword. In *The Arcades Project* by W. Benjamin, ix–xiv. Cambridge, MA: Belknap Press of Harvard University Press.

Einstein, A., and L. Infeld. 1966. *The Evolution of Physics: From Early Concepts to Relativity and Quanta*. New York: Simon and Schuster.

Engels, F. 1958. *The Condition of the Working Class in England*. Trans. W. O. Henderson and W. H. Chaloner. Oxford: Basil Blackwell.

———. 1979. *The Housing Question*. Moscow: Progress.

Esslin, M. 1980. *Brecht, a Choice of Evils: A Critical Study of the Man, His Work, and His Opinions*. London: Eyre Methuen.

Fabian, J. 1983. *Time and the Other: How Anthropology Makes Its Object*. New York: Columbia University Press.

Fan, C. C. 2011. Settlement intention and split households: Findings from a survey of migrants in Beijing's urban villages. *China Review* 11 (2): 11–42.

Fan, C. C., M. Sun, and S. Zheng. 2011. Migration and split households: A comparison of sole, couple, and family migrants in Beijing, China. *Environment and Planning A* 43 (9): 2164–85.

Fan, C. C., and W. W. Wang. 2008. The household as security: Strategies of rural-urban migrants in China. In *Migration and Social Protection in China*, ed. I. Nielsen and R. Smyth, 205–43. Hackensack, NJ: World Scientific.

Felski, R. 2000. The invention of everyday life. *New Formations* 39:15–31.

Fernandes, E. 2007. Constructing the "right to the city" in Brazil. *Social Legal Studies* 16 (21): 201–19.

Fernandez, R. 2003. *Mappers of Society: The Lives, Times, and Legacies of Great Sociologists*. Westport, CT: Praeger.

Fliegelman, J. 1993. *Declaring Independence: Jefferson, Natural Language, & the Culture of Performance*. Stanford, CA: Stanford University Press.

Foucault, M. 1980. *Power/Knowledge: Selected Interviews & Other Writings, 1972–1977*. Ed. C. Gordon. Trans. A. Fontana and P. Pasquino. New York: Pantheon Books.

———. 1995. *Discipline and Punish: The Birth of the Prison*. Trans. by A. Sheridan. New York: Vintage Books.

———. 2003. *"Society Must Be Defended": Lectures at the Collège de France, 1975–76*. Trans. D. Macey. New York: Picador.

———. 2004. *The Birth of Biopolitics: Lectures at the Collège de France, 1978–79*. Ed. M. Senellart. Trans. G. Burchell. Houndmills, Basingstoke, Hampshire, UK: Palgrave Macmillan.

———. 2007a. The meshes of power. Trans. G. Moore. In *Space, Knowledge, and Power: Foucault and Geography*, ed. J. W. Crampton and S. Elden, 153–62. Farnham, Surry, UK: Ashgate.

———. 2007b. *Security, Territory, Population: Lectures at the Collège de France, 1977–1978*. Ed. M. Senellart. Trans. G. Burchell. New York: Picador.

Galison, P. 2003. *Einstein's Clocks, Poincaré's Maps: Empires of Time*. New York: W. W. Norton.

Gaukroger, S. 1995. *Descartes: An Intellectual Biography*. Oxford: Clarendon Press, Oxford University Press.

Ghannam, F. 2008. Two dreams in a global city: Class and space in urban Egypt. In *Other Cities, Other Worlds: Urban Imaginaries in a Globalizing Age*, ed. A. Huyssen, 267–88. Durham, NC: Duke University Press.

Gibson, J. J. 1966. *The Senses Considered as Perceptual Systems*. Boston: Houghton Mifflin.

———. 1979. *The Ecological Approach to Visual Perception*. Boston: Houghton Mifflin.

Giddens, A. 1971. *Capitalism and Modern Social Theory: An Analysis of the Writings of Marx, Durkheim, and Max Weber*. Cambridge, UK: Cambridge University Press.

———. 1984. *The Constitution of Society: Outline of the Theory of Structuration*. Berkeley: University of California Press.

———. 1985. *The Nation-State and Violence*. Vol. 2 of *A Contemporary Critique of Historical Materialism*. Berkeley: University of California Press.

———. 1990. *The Consequences of Modernity*. Stanford, CA: Stanford University Press.

Goffman, E. 1956. Embarrassment and social organization. *American Journal of Sociology* 62(3): 264–71.

———. 1961. *Asylums: Essays on the Social Situation of Mental Patients and Other Inmates*. Chicago: Aldine.

———. 1963. *Behavior in Public Places: Notes on the Social Organization of Gatherings*. New York: Free Press.

———. 1969. *Strategic Interaction*. Philadelphia: University of Pennsylvania Press.

———. 1971. *Relations in Public: Microstudies of the Public Order*. New York: Basic Books.

———. 1972. The neglected situation. In *Language and Social Context: Selected Readings*, ed. P. P. Giglioli, 61–66. Harmondsworth, Middlesex, UK: Penguin Books.

———. 1974. *Frame Analysis: An Essay on the Organization of Experience*. Cambridge, MA: Harvard University Press.

———. 1983. The interaction order: American Sociological Association, 1982 Presidential Address. *American Sociological Review* 48 (1): 1–17.

Gottdiener, M. 1985. *The Social Production of Urban Space*. Austin: University of Texas Press.

Greenfield, A. 2006. *Everyware: The Dawning Age of Ubiquitous Computing*. Berkeley, CA: New Riders.

Greenstone, M., and A. Looney. 2011. Trends. *Milkin Institute Report*, Third Quarter. brookings.edu/~/media/Files/rc/papers/2011/07_milken_greenstone_looney/07_milken_greenstone_looney.pdf.

Grosz, E. 1994. *Volatile Bodies: Toward a Corporeal Feminism*. Bloomington: Indiana University Press.

———. 1999. Merleau-Ponty and Irigaray in the flesh. In *Merleau-Ponty, Interiority and Exteriority, Psychic Life and the World*, ed. D. Olkowski and J. Morley, 145–66. Albany: State University of New York Press.

Hacking, I. 2004. Between Michel Foucault and Erving Goffman: Between discourse in the abstract and face-to-face interaction. *Economy and Society* 33 (3): 277–302.

Hägerstrand, T. 1967. *Innovation Diffusion as a Spatial Process*. Trans. A. Pred, with the assistance of G. Haag. Chicago: University of Chicago Press.

———. 1970. What about people in regional science? *Papers of the Regional Science Association* 24:7–21.

———. 1982. Diorama, path, and project. *Tijdschrift voor Econ. En Soc. Geografie* 73 (6): 323–39.

———. 2004. The two vistas. Trans. Tommy Carlstein. *Geografiska Annaler* 86 (4): 315–23.

Harney, M. 2007. Merleau-Ponty, ecology, and biosemiotics. In *Merleau-Ponty and Environmental Philosophy: Dwelling in the Landscapes of Thought*, ed. S. L. Cataldi and W. S. Hamrick, 133–46. Albany: State University of New York Press.

Harrison, B., and B. Bluestone. 1988. *The Great U-Turn: Corporate Restructuring and the Polarizing of America*. New York: Basic Books.

Harvey, D. 1974. *The Political Economy of Urbanization in Advanced Capitalist Societies: The Case of the United States*. Baltimore: Johns Hopkins University, Center for Metropolitan Planning and Research.

———. 1985. *The Urbanization of Capital: Studies in the History and Theory of Capitalist Urbanization*. Baltimore: Johns Hopkins University Press.

——. 1989. *The Condition of Postmodernity: An Enquiry into the Origins of Cultural Change*. Oxford: Basil Blackwell.

——. 1995. *The Limits to Capital*. London: Verso.

——. 2002. The urban process under capitalism: A framework for analysis. In *The Blackwell City Reader*, ed. G. Bridge and S. Watson, 116–24. Malden, MA: Blackwell.

——. 2010. *The Enigma of Capital and the Crises of Capitalism*. Oxford: Oxford University Press.

Hasek, J. 1974. *The Good Soldier Svejk and His Fortunes in the World War*. Trans. C. Parrott. New York: Thomas Y. Crowell.

Hegel, G. W. F. 1977. *Phenomenology of Spirit*. Trans. A. V. Miller. Oxford: Oxford University Press.

Heidegger, M. 1996. *Being and Time*. Trans. J. Stambaugh. Albany: State University of New York Press.

Herzfeld, M. 2005. *Cultural Intimacy: Social Politics in the Nation-State*. New York: Routledge.

Hesiod. 1988. *Theogony and Works and Days*. Trans. M. I. West. New York: Oxford University Press.

Highmore, B. 2006. *Michel de Certeau: Analyzing Culture*. London: Continuum.

Hollander, S. 2008. *The Economics of Karl Marx: Analysis and Application*. New York: Cambridge University Press.

Holt-Jensen, A. 1999. *Geography: History and Concepts, a Students' Guide*. Los Angeles: Sage.

Hsing, Y.-T. 2010. *The Great Urban Transformation: Politics of Land and Property in China*. Oxford: Oxford University Press.

Hume, D. 1978. *A Treatise of Human Nature*. Oxford: Oxford University Press.

Husserl, E. 1964. *The Paris Lectures*. Trans. P. Koestenbaum. The Hague: Martinus Nijhoff.

——. 1999. *Cartesian Meditations: An Introduction to Phenomenology*. Trans. D. Cairns. Dordrecht: Kluwer Academic.

——. 2002. The world of the living present and the constitution of the surrounding world that is outside the flesh. In *Husserl at the Limits of Phenomenology, Including Texts by Edmund Husserl*, by M. Merleau-Ponty, 132–54. Ed. L. Lawlor with B. Bergo. Evanston, IL: Northwestern University Press.

Iyer, P. 2004. *Sun after Dark: Flights into the Foreign*. New York: Alfred A. Knopf.

Jameson, F. 2003. The end of temporality. *Critical Inquiry* 29 (4): 695–718.

Jammer, M. 1954. *Concepts of Space: the History of Theories of Space in Physics*. Cambridge, MA: Harvard University Press.

Janik, A., and S. Toulmin. 1973. *Wittgenstein's Vienna*. New York: Simon and Schuster.

Johnston, R. J. 1980. *City and Society*. Harmondsworth, UK: Penguin.

Kaku, M. 2004. *Einstein's Cosmos: How Albert Einstein's Vision Transformed Our Understanding of Space and Time*. New York: W. W. Norton.

Knox, P., J. Agnew, and L. McCarthy. 2008. *The Geography of the World Economy*. London: Hodder Education.

Kraul, C. 2012. Latin American air travel soars. *Los Angeles Times*, 9 March, B1, 6.

Krugman, P. 1995. *The Age of Diminished Expectations: U.S. Economic Policy in the 1990s*. Cambridge, MA: MIT Press.

Kwan, M. 2004. GIS methods in time-geographic research: Geocomputation and geovisualization of human activity patterns. *Geografiska Annaler* 86 (4): 267–80.

Lakoff, G., and M. Johnson. 1999. *Philosophy in the Flesh: The Embodied Mind and Its Challenge to Western Thought*. New York: Basic Books.

Larsen, S. C., and J. T. Johnson. 2012. Toward an open sense of place: Phenomenology, affinity, and the question of being. *Annals of the Association of American Geographers* 102 (3): 632–46.

Latouche, S. 1996. *The Westernization of the World: The Significance, Scope, and Limits of the Drive towards Global Uniformity*. Trans. R. Morris. Cambridge, UK: Polity Press.

Laughlin, R. B. 2005. *A Different Universe: Reinventing Physics from the Bottom Down*. New York: Basic Books.

Lee, C. K. 2007. Mapping the terrain of Chinese labor ethnography. In *Working in China: Ethnographies of Labor and Workplace Transformation*, ed. C. K. Lee, 1–12. London: Routledge.

Lee, D. 2011. As U.S. stumbles, companies invest in consumer growth overseas. *Los Angeles Times*, 8 August.

Lefebvre, H. 1976. Reflections on the politics of space. *Antipode* 8 (2): 30–37.

———. 1979. Space: Social product and use value. Trans. J. W. Freiberg. In *Critical Sociology: European Perspectives*, ed. J. W. Freiberg, 285–95. New York: Irvington.

———. 1991. *The Production of Space*. Trans. D. Nicholson-Smith. Oxford: Blackwell.

Leibniz, G. W., and S. Clarke. 1956. *The Leibniz-Clarke Correspondence, Together with Extracts from Newton's Principia and Optiks*, ed. H. G. Alexander. Manchester, UK: Manchester University Press.

Levins, R., and R. Lewontin. 1985. *The Dialectical Biologist*. Cambridge, MA: Harvard University Press.

Lifescapes International. Don Brinkerhoff honored in China. http://www.lifescapesintl.com/news-events-orchid-garden/.

Lin, G. C. S. 2007. Reproducing spaces of Chinese urbanization: New city-based and land-centered urban transformation. *Urban Studies* 44 (9): 1828–55.

Lingis, A. 1968. Translator's preface. In *The Visible and the Invisible* by M. Merleau-Ponty, xl–lvi. Evanston, IL: Northwestern University Press.

Livingstone, D. N. 2003. *Putting Science in Its Place: Geographies of Scientific Knowledge*. Chicago: University of Chicago Press.

Lyall, S. 2011. Seeking space, well-to-do Londoners dig deep. *New York Times*, 31 August.

Lyon, J. K. 1980. *Bertolt Brecht in America*. Princeton, NJ: Princeton University Press.

Malpas, J. E. 1999. *Place and Experience: A Philosophical Topography*. Cambridge, UK: Cambridge University Press.

Malthus, T. R. *On Population*. Ed. G. Himmelfarb. New York: Modern Library, 1960.

Mandeville, B. 1728. *The Fable of the Bees*. 5th ed. London.

Manheim, R., and J. Willett. 1975. Introduction. In *Brecht: Collected Plays*, vol. 7, ed. R. Manheim and J. Willett. New York: Vintage Books.

Manning, P. 1992. *Erving Goffman and Modern Sociology*. Stanford, CA: Stanford University Press.

Martensson, S. 1979. *On the Formation of Biographies in Space-Time Environments*. Lund Studies in Geography, no. 47. Royal University of Lund, Department of Geography, GWK Gleerup.

Martin, F. D. 1981. *Sculpture and Enlivened Space: Aesthetics and History*. Lexington: University Press of Kentucky.

Marx, K. 1994. *Karl Marx: Selected Writings*. Ed. L. H. Simon. Trans. L. D. Easton and K. H. Guddat. Indianapolis: Hackett.

———. 1996. *Capital*, vol. 1. In *Collected Works* by Karl Marx and Frederick Engels. Vol. 35. New York: International.

———. 1997. *Capital*, vol. 2. In *Collected Works* by Karl Marx and Frederick Engels. Vol. 36. New York: International.

Massey, D. 1992. Politics and space/time. *New Left Review* 196:65–84.

———. 1994. *Space, Place, and Gender*. Minneapolis: University of Minnesota Press.

———. 2005. *For Space*. London: Sage.

———. 2007. *World City*. Cambridge, UK: Polity Press.

Massumi, B. 1988. Notes on the translation and acknowledgments. In *A Thousand Plateaus: Capitalism and Schizophrenia*, by G. Deleuze and F. Guattari, xvi–xix. London: Althone Press.

Mauss, M. 1967. *The Gift: Forms and Functions of Exchange in Archaic Societies*. Trans. I. Cunnison. New York: W. W. Norton.

McCleary, R. C. 1964. Translator's preface. In *Signs* by M. Merleau-Ponty, ix–xxxii. Evanston, IL: Northwestern University Press.

Merleau-Ponty, M. 1962. *Phenomenology of Perception*. Trans. C. Smith. London: Routledge & Kegan Paul.

———. 2004. *The World of Perception*. Trans. by O. Davis. London: Routledge.

———. 2010. *Institution and Passivity: Course Notes from the Collège de France (1954–1955)*. Trans. L. Lawlor and H. Massey. Evanston, IL: Northwestern University Press.

Minkowski, Hermann. Space and time. In *Problems of Space and Time*, ed. J. J. C. Smart, 297–312. New York: Macmillan, 1964.

More bleak job numbers. 2011. *New York Times* editorial. 7 October.

Myerson, H. 2011. The fallacy of post-industrial prosperity. *Washington Post*, 4 September.

Nigerian Students for Environmental Action. 2010. Environmental destruction and human rights in the Niger Dela (Ogochukwu). *Class Blog*, 29 November.

Nnanna, O. 2009. Confronting Tompolo. *Vanguard*, 18 June. vanguardngr.com/2009/06/confronting-tompolo/.

Noë, A. 2004. *Action in Perception*. Cambridge, MA: MIT Press.

———. 2009. *Out of Our Heads: Why You Are Not Your Brain, and Other Lessons from the Biology of Consciousness*. New York: Hill and Wang.

Nolan, P., and J. Zhang. 2010. Global competition after the financial crisis. *New Left Review* 64 (July/August): 97–108.

Nugent, J. B., and J. A. Robinson. 2010. Are factor endowments fate? *Journal of Iberian and Latin American Economic History* 28 (1): 45–82.

Olkowski, D. 1999. Introduction: The continuum of interiority and exteriority in the thought of Merleau-Ponty. In *Interiority and Exteriority, Psychic Life and the World*, by M. Merleau-Ponty, 1–21. Ed. D. Olkowski and J. Morley. Albany: State University of New York Press.

Palm, R., and A. Pred. 1974. A time-geographic perspective on problems of inequality for women. Working paper. Berkeley: Institute of Urban and Regional Development, University of California.

Parrott, C. 1974. Introduction. In *The Good Soldier Svejk and His Fortunes in the World War*, by J. Hasek. New York: Thomas Y. Crowell.

Paterson, M. 2007. *The Senses of Touch: Haptics, Affects and Technologies*. Oxford: Berg.

———. 2009. Haptic geographies: Ethnography, haptic knowledges and sensuous dispositions. *Progress in Human Geography* 33 (6): 766–88.

Paul, G. A. 1956. Wittgenstein. In *The Revolution in Philosophy*, ed. G. Ryle. London: Macmillan.

Philo, C. 2007. "Bellicose history" and "local discurvities": An archeological reading of Michel Foucault's *Society Must Be Defended*. In *Space, Knowledge, and Power: Foucault and Geography*, ed. J. W. Crampton and S. Elden, 341–67. Farnham, Surry, UK: Ashgate.

Poggi, G. 1972. *Images of Society: Essays on the Sociological Theories of Tocqueville, Marx, and Durkheim*. Stanford, CA: Stanford University Press.

Poincaré, Henri. 1963. Space and time. In *Mathematics and Science: Last Essays*, trans. John W. Bolduc. New York: Dover.

Popper, K. 1944. The poverty of historicism. Part 2: A criticism of historicist methods. *Economica* 2 (43): 119–37.

Poster, M. 1992. The question of agency: Michel de Certeau and the history of consumerism. *Diacritics* 22 (2): 94–107.

Pratt, J. H. 1942. American prime meridians. *Geographical Review* 32 (2): 233–44.

Pred, A. 1977. The choreography of existence: Comments on Hägerstrand's time-geography and its usefulness. *Economic Geography* 53 (2): 207–21.

Protevi, J. 2009. *Political Affect: Connecting the Social and the Somatic*. Minneapolis: University of Minnesota Press.

Robb, G. 2010. The Divine Sarah. *New York Review of Books* 57 (15): 8–12.

Rose, G. 1993. *Feminism and Geography: The Limits of Geographical Knowledge*. Cambridge, UK: Polity Press.

Rose, N. 1996. Governing "advanced" liberal democracies. In *Foucault and Political Reason: Liberalism, Neo-liberalism and Rationalities of Government*, ed. A. Barry, T. Osborne, and N. Rose, 37–64. Chicago: University of Chicago Press.

Rosenblum, L. D. 2010. *See What I'm Saying: The Extraordinary Powers of Our Five Senses*. New York: W. W. Norton.

Roth, A. 2011. Environmental destruction and human rights in the Niger Delta. *Freedom from Fear Magazine*, 2 November.

Rouse, J. 1996. *Engaging Science: How to Understand Its Practices Philosophically*. Ithaca, NY: Cornell University Press.

Sack, R. 1997. *Homo Geographicus*. Baltimore: Johns Hopkins University Press.

Sacks, O. W. 2003. A neurologist's notebook: The mind's eye—what the blind see. *New Yorker*, 28 July: 48–59.

Sagan, D. 2010. Introduction: Umwelt after Uexküll. In *A Foray into the World of Animals and Humans, with A Theory of Meaning*, by J. von Uexküll, 1–34. Trans. J. D. O'Neil. Minneapolis: University of Minnesota Press, 1–34.

Sambursky, S. 1962. *The Physical World of Late Antiquity*. London: Routledge and Kegan Paul.

Sartre, J.-P. 1950. *The Diary of Antoine Roquentin*. Trans. from the French *La Nausée* by L. Alexander. London: John Lehmann.

Saslow, E. 2011. Virginia house painter fights to keep business as recession becomes a way of life. *Washington Post*, 29 August.

Sassen, S. 2002. Locating cities on global circuits. *Environment and Urbanization* 14 (1): 13–30.

Scheper-Hughes, N. 2005. The last commodity: Post-human ethics and the global traffic in "fresh" organs. In *Global Assemblages: Technology, Politics, and Ethics as Anthropological Problems*, ed. A. Ong and S. J. Collier, 145–68. Malden, MA: Blackwell.

Schutz, A. 1967. *The Phenomenology of the Social World*. Trans. G. Walsh and F. Lehnert. Evanston, IL: Northwestern University Press.

Schwanen, T. 2007. Matter(s) of interest: Artefacts, spacing and timing. *Geografiska Annaler* 89 (1): 9–22.

Scott, J. C. 1986. Everyday forms of peasant resistance. In *Everyday Forms of Peasant Resistance in South-East Asia*, ed. J. C. Scott and B. J. Tria Kerkvliet, 5–35. London: Frank Cass.

Sewell, W. H. 2005. *Logics of History: Social Theory and Social Transformation*. Chicago: University of Chicago Press.

Shakespeare, W. 1936. *Second Part of King Henry the Fourth*. In *The Complete Works of Shakespeare*. Ed. George Lyman Kittredge. Boston: Ginn and Co.

Sheldon, R., and E. Arens. 2005. Make it snuggle in the palm: The commodification of touch. In *The Book of Touch*, ed. C. Classen, 426–28. Oxford: Berg.

Shell Nigeria. 2011. More sabotage spills in Niger Delta since SPDC shut down Imo River production. Press release, 11 January. http://www.shell.com.ng/media /2011-media-releases/more-sabotage-spills.html.

Silverstone, R. 1994. *Television and Everyday Life*. London: Routledge.

Soja, E. W. 2010. *Seeking Spatial Justice*. Minneapolis: University of Minnesota Press.

Straus, E. 1963. *The Primary World of Senses: A Vindication of Sensory Experience*. Trans. J. Needleman. New York: Free Press.

Strong, H. M. 1935. Universal world time. *Geographical Review* 25:479–84.

Sullivan, R. 2011. *Geography Speaks: Performative Aspects of Geography*. Surrey, UK: Ashgate.

Thompson, E. P. 1963. *The Making of the English Working Class*. New York: Pantheon Books.

Tocqueville, A. de. 1955. *The Old Regime and the French Revolution*. Trans. S. Gilbert. Garden City, NY: Doubleday Anchor Books.

———. 1969. *Democracy in America*. Trans. G. Lawrence. Ed. J. P. Mayer. Garden City, NY: Anchor Books.

Touraine, A. 1988. *Return of the Actor: Social Theory in Postindustrial Society*. Trans. M. Godzich. Minneapolis: University of Minnesota Press.

Tuan, Y.-F. 1977. *Space and Place: The Perspective of Experience*. Minneapolis: University of Minnesota Press.

———. 1980. Rootedness versus sense of place. *Landscape* 24 (1): 3–8.

Uexküll, J. de. 2010. *A Foray into the World of Animals and Humans, with A Theory of Meaning*. Trans. J. D. O'Neil. Minneapolis: University of Minnesota Press.

Van Loon, H. W. 1932. *Van Loon's Geography: The Story of the World We Live In*. New York: Simon and Schuster.

Verbücken, M. 2011. Towards a new sensorality. In *The New Everyday: Views on Ambient Intelligence*, ed. E. Aarts and S. Marzano, 54–59. Rotterdam: 010 Publishers.

Vidal, J. 2010. Nigeria's agony dwarfs the Gulf oil spill. The US and Europe ignore it. *Guardian/Observer* (London), 29 May.

Virilio, P. 1993. The primal accident. In *The Politics of Everyday Fear*, ed. B. Massumi, 211–18. Minneapolis: University of Minnesota Press.

Watts, M. 2000. *Struggles over Geography: Violence, Freedom, and Development at the Millennium*. Hettner-Lectures no. 3, 1999. Heidelberg, Germany: Department of Geography, University of Heidelberg.

———. 2009. Crude politics: Life and death on the Nigerian oil fields. Working Paper no. 25, Niger Delta: Economies of Violence Project. Berkeley: Institute of International Studies, University of California.

Weber, M. 1959. *From Max Weber: Essays in Sociology*. Trans. and ed. H. H. Gerth and C. W. Mills. New York: Oxford University Press.

———. 1968. *Economy and Society: An Outline of Interpretive Sociology*. Vol. 3. Ed. G. Roth and C. Wittich. New York: Bedminster Press.

Wheeler, W. 2006. *The Whole Creature: Complexity, biosemiotics and the evolution of culture*. London: Lawrence & Wishart.

White, C. P. 1986. Everyday resistance, socialist revolution and rural development: The Vietnam case. In *Everyday Forms of Peasant Resistance in South-East Asia*, ed. J. C. Scott and B. J. Tria Kerkvliet, 49–63. London: Frank Cass.

Willbros. 2000. Brochure. willbros.com/fw/filemanager/fm_file_manager _download.asp?.

Williams, R. 1973. *The Country and the City*. Oxford: Oxford University Press.

Williams, R. G. 1994. *State and Social Evolution: Coffee and the Rise of National Governments in Central America*. Chapel Hill: University of North Carolina Press.

Wittgenstein, L. 2001. *Philosophical Investigations*. Trans. G. E. M. Anscombe. Malden, MA: Blackwell.

Wolch, J., and G. DeVerteuil. 2001. New landscapes of urban property management. In *Timespace: Geographies of Temporality*, ed. J. May and N. Thrift, 149–68. London: Routledge.

Wolpert, S., and D. Menon. 2011. Sound and vision work hand in hand, UCLA psychologists report. UCLA Newsroom, 8 December. http://newsroom.ucla.edu /releases/sound-and-vision-work-hand-in-220261.

Yergin, D. 1991. *The Prize: The Epic Quest for Oil, Money, and Power*. New York: Simon and Schuster.

Young, I. M. 1990. Throwing like a girl: A phenomenology of feminine body comportment, motility, and spatiality. In *The Thinking Muse: Feminism and Modern French Philosophy*, ed. J. Allen and I. M. Young, 51–70. Bloomington: Indiana University Press.

Zheng, S., F. Long, C. Fan, and Y. Gu. 2009. Urban villages in China: A 2008 survey of migrant settlements in Beijing. *Eurasian Geography and Economics* 50 (4): 425–46.

Zoellick, R. 2011. The big questions China still has to answer. *Financial Times* (USA), 2 September, 9.

Zoulas, J. G., and A. R. Orme. 2007. Multidecadal-scale beach changes in the Zuma littoral cell, California. *Physical Geography* 28 (4): 277–300.

INDEX

de Mandeville, Bernard, 99–100
Descartes, René, 50, 153, 157–59, 170–71, 179
DeVerteuil, Geoffrey, 35
Dionne, E. J., 123

Egypt, 101, 118, 124, 126–27, 175–76
Eickelman, Dale F., 60
Eiland, Howard, 163
Einstein, Albert, 177; on gravitational red shift, 55; and Newton, compared to, 38; on relativity, 42–43, 45, 47, 51–52; on spacetime, 38, 42–43, 47, 51–52
Engels, Frederick: and China's urban villages, 105; on the middle class, 116, 118–19; on the ruling class, 125; on the working class, 103; —, American, 122–24; —, English, 98, 101–2
Esslin, Martin, 70–71, 73
Europe: BP's headquarters in, 91; and Brecht, 69, 74; and English "folk," 102; and the EU, economic issues in, 121; and the post-colonial, 127–29; Shell's headquarters in, 91

Fabian, Johannes, 40
Fan, C. Cindy, 105–8, 110–11
Felski, Rita, 29–30
Fernandes, Edésio, 79
Fernandez, Ronald, 114
Fliegelman, Jay, 114, 179
Foucault, Michel, 177; and de Certeau, compared to, 61–62; and Goffman, 3–5, 176; on migration, 106; and the milieu, 8, 15–25, 28, 175

Galison, Peter, 25, 52
Gaukroger, Stephen, 50, 179
Ghannam, Farha, 101, 121, 126–27, 179
Gibson, J. J.: and the environment and the mind, 142–43; on the haptic system, 165; and indexical coordination, 136; on the visual system, 147, 149–50, 164
Giddens, Anthony: on Foucault, 4; on Goffman, 4, 15, 28; on Marx and history, 94; on place, 52–53; on the reserve army, 99
Goffman, Erving, 2; and Foucault, compared to, 3–5, 25, 61; and malingering, 66; and the situation, 8–15, 175–76; —, compared with the milieu, 21, 23–24; and time-geography, 28, 31–32

Gottdiener, Mark, 78–79
Greenfield, Adam, 152
Greenstone, Michael, 123
Grosz, Elizabeth, 134–35, 138, 140, 142
Gu, Yizhen, 105
Guattari, Félix, 16, 83–84

Hacking, Ian: and comparison of Goffman and Foucault, 3–5, 8, 175–76; on Foucault, 22–23, 25; and Poster, Mark, 61
Hägerstrand, Torsten, 26–35, 38–39, 179
haptic, the, 145, 151–53, 165–66
Harney, Maurita, 136
Harrison, Bennett, 112
Harvey, David: and China, 109; and economic reproduction, 95, 125; on enclosure, 100; on multiplier effects, 97; on neoliberalism, 102; and time-geography, 36
Hasek, Jaroslav, 59, 63–67, 70, 76, 92
Hegel, Georg Wilhelm, 77–78, 135, 140–41
Heidegger, Martin, 40
Herzfeld, Michael, 64
Hesiod, 50, 177
Highmore, Ben, 62, 64, 67
Hollander, Samuel, 99
Holt-Jensen, Arild, 44
Hsing, You-tien, 107–10
Hume, David, 148
Husserl, Edmund, 133–34, 145, 148, 170–71

India, 6, 114, 121
Infeld, Leopold, 45
Iyer, Pico, 128

Jameson, Frederic, 36–38
Jammer, Max, 43
Janik, Allan, 43
Johnson, Mark, 153–54
Johnston, Ron, 96

Kaku, Michio, 45, 55–56
Knox, Paul, 96
Kraul, Chris, 120–21
Krugman, Paul, 112
Kwan, Mei-Po, 39

Lakoff, George, 153–54
Latouche, Serge, 46
Laughlin, Robert B., 56
Lee, Don, 108, 120–21

ruse, the (*continued*)
    and Hasek, 63–64; and history, 76; and
    Schweyk, 66–69; and Svejk, 64–66
Russia, 114, 123

Sack, Robert, 53, 131–32
Sacks, Oliver, 146
Sagan, Dorion, 148, 159–60
Sambursky, Samuel, 42
Sartre, Jean-Paul, 40, 149, 155
Saslow, Eli, 123–24
Sassen, Saskia, 19
Scheper-Hughes, Nancy, 91
Schutz, Alfred, 136–39
Schwanen, Tim, 39
Schweyk (Brecht character), 59, 63, 67–69,
    75–76, 92
Scott, James C., 75–76
Sewell, William H., 7
Shakespeare, William, 126
Silverstone, Roger, 59
Soja, Edward W., 41, 47–49, 103–4
spaceplacetime, 50–51
spacetime: difficulty conceiving, 43; and
    Einstein, 38, 42–43, 45, 47, 51–52; and
    Hägerstrand, 27; and Hesiod, 52; lip
    service to, 38; recapitulation of, 176; and
    spaceplacetime, 50, 58; taking the term
    literally, 41
spatial secretion, 91–93; and Lefebvre, 6, 59,
    77; and Nigerian oil complex, 80, 87–88
Straus, Erwin, 142–43, 157
Stripling, Robert, 71–75
Strong, Helen M., 46
Sun, Mingjie, 111
Svejk (Hasek character), 63–66, 75–76, 92
Sweden, 32–33, 38

Taine, Hippolyte, 16
Thomas, J. Parnell, 70, 74–75
Thompson, Edward P., 100
time-geography, 26–28, 31, 33–36, 38–40

Tocqueville, Alexis de, 116–17, 119
Toulmin, Stephen, 43
Touraine, Alain, 4
Tuan, Yi-Fu, 53, 131–32

United States: and Brecht, 66, 69–70, 74; and
    class and aesthetics, 127–29; demographics
    of, 110–11; and eating habits, 1; and
    Goffman, 11; housing in Southwest of,
    97; and immigration patterns, 106; and
    the middle class, 6, 111–14, 116–21; and
    neoliberalism, 103; and Nigerian oil, 80;
    and organ transplants, 92; and private
    property, 88; and space, time, and place, 55

van Loon, Hendrik Willem, 46
Verbücken, Marion, 152
Vidal, John, 80
Virilio, Paul, 17
von Uexküll, Jacob, 51, 159, 171

Wang, Wenfei Winnie, 106–8
Watts, Michael, 6, 80–81, 85–90, 122–23, 179
Weber, Max, 116–19, 161–62
Welch, Joseph, 74–75
Wheeler, Wendy, 165
White, Christine Pelzer, 76
Willbros Group, 84–85
Willet, John, 59
Williams, Raymond, 101
Wittgenstein, Ludwig, 164, 172
Wolch, Jennifer, 35
Wolpert, Stuart, 145

Yergin, Daniel, 82, 87
Young, Iris Marion, 134, 138, 140–41

Zhang, Jin, 122
Zheng, Siqi, 104–6, 111
Zoellick, Robert, 108
Zoulas, James G., 57–58

||||||| ||||| ||| |||||| |||| ||||| ||||| ||| |||

www.ingramcontent.com/pod-product-compliance
Lightning Source LLC
Chambersburg PA
CBHW010114270326
41928CB00021B/3253

* 9 780820 351674 *